THE GEOGRAPHY COLLECTION

ENERGY MATTERS

Rosemary Hector

Series Editor
Ian Selmes

Hodder & Stoughton
A MEMBER OF THE HODDER HEADLINE GROUP

ACKNOWLEDGEMENTS

The author and publishers would like to thank the following for permission to reproduce materials in this book. Every effort has been made to trace and acknowledge all copyright holders, but if any have been overlooked the publishers will be pleased to make the necessary arrangements.

British Gas, Figure 2.5; Coal News, Figure 2.1; BP Statistical Review of World Energy, Figures 1.1, 1.2, 1.4, 1.5, 2.2; AG Hector; Dr MF Hector, Figure 4.6; *The Ecologist*; Heinemann Educational Books (Nigeria) Plc, *Hidesong*, Aigboje Higo, page 35; © *The Independent*, Figures 2.9, 3.3; HMSO, *Digest of UK Energy Statistics*, Figure 3.4; James Rose, assistance in Section 2; The Japanese Embassy, Figures 3.5, 3.6; The National Grid Company plc; Natural Resources Canada, information in DME; Nigerian National Petroleum Corporation, Figure 4.1; Honourable Jake Epp for his speech, page 44; Uranium Institute, Figure 1.3; D Whaley, Bristol University Library; Scottish National Party, Figure 2.4; *The Scotsman*, Figure 1.6; Dr Emilie Smith, assistance in Section 4; Dr BD Titus, Canadian Forestry Service, pages 41–42; World Wide Fund for Nature and World Conservation and Monitoring Service, Figure 4.5.

The publishers would also like to thank the following for giving permission to reproduce copyright photographs in this book.

Kippa Matthews, Figure 1.7; Figure 2.11 by courtesy of TNC Consulting AG, ©1995 Männedorf Switzerland. Cover photo: lightning over the Grand Canyon, Tony Stone Images.

All other photos belong to the author.

Inside artwork by Jeff Edwards.

ISBN 0 340 61879 5

First published 1995
Impression number   10 9 8 7 6 5 4 3 2 1
Year   1999 1998 1997 1996 1995

Copyright © 1995 Rosemary Hector

All rights reserved. No part of this publication may be reproduced or transmitted in any form or by any means, electronic or mechanical, including photocopy, recording, or any information storage and retrieval system, without permission in writing from the publisher or under licence from the Copyright Licensing Agency Limited. Further details of such licences (for reprographic reproduction) may be obtained from the Copyright Licensing Agency Limited, of 90 Tottenham Court Road, London W1P 9HE.

Typeset by Wearset, Boldon, Tyne and Wear.
Printed in Great Britain for Hodder & Stoughton Educational, a division of Hodder Headline Plc, 338 Euston Road, London NW1 3BY by Redwood Books, Trowbridge, Wilts.

contents

Section 1	**The nature of energy**	page 4
	Global energy patterns; Energy issues at global, regional and local scales	
Section 2	**Releasing energy**	page 12
	Fossil fuels; Nuclear power; Renewable energy	
Section 3	**Harnessing energy: the UK and Japan**	page 26
	Energy policy in the UK: grey or green? Japan's catch-up capitalism	
Section 4	**Africa: energy rich and energy poor**	page 33
	Nigeria and the Sinatra doctrine; East Africa: an energy challenge	

Decision making exercise: an energy policy for Newfoundland — page 40

Project suggestions — page 45

Glossary — page 46

References and useful addresses — page 47

Index — page 48

thematic contents

Transport
oil (15), LPG (18) electricity grid (20), Nigerian oil (33), (34), East African energy (37).

Trade
in oil, petro-dollars (7), Japan's imports (30), Nigerian oil (33), Nigerian coal (35).

Hazards
third world debt (8), Greenhouse Effect (8), sea levels (8), acid rain (9), coal mining (14), oil spills (18), Uranium (19), Chernobyl (20), nuclear waste (20), dash for gas (27), Japanese pollution (31), radioactivity (34), firewood clearances (37).

Green
alternative resources (6), biomass (6), (24), afforestation (8), Greenhouse Effect (8), visual pollution (9), Clean Air Acts (14), renewable energy (21), anaerobic digestion (24), gas turbines (27), the wind rush (27), Japanese HEP (30).

Political
OPEC (7), IPCC (8), NCB (14), Coal Industry Act (14), storing radioactive materials (20), US National Energy Strategy (24), UK energy strategy (26), Japanese policies (30), East African policies (39), Newfoundland policy (40), Hibernia Development Project (44).

Skills
Spearman Rank Correlation (6), spot sampling (11), project suggestions (45).

the geography collection is a series of texts designed for students of the new A-level geography syllabuses and has been written by a team of eminent geography teachers. The series provides an enlightened approach to the discipline, opening up the world of the geographer through contemporary text, clarifying illustrations, integrated questions and decision making exercises. The Core Text, *World Wide*, and Option Booklets contain guidelines for geographical skills and project work together with a glossary of important terms. A feature of the series are the themes that run through each book, signalled by icons, giving extra flexibility in the way that students can use the texts.

SECTION 1

The nature of energy

KEY IDEAS

- The wealthier a country, the higher its energy consumption and the more diverse its energy sources. The poorest countries are most heavily dependent on 'people power' and biomass, both of which are difficult to measure.
- Energy sources are unevenly distributed. This affects global power relationships.
- Energy uses have an impact at different scales, and affect economic and ecological systems.
- The energy choices and patterns a society creates are related to the political, social and economic patterns in that society.

GLOBAL ENERGY PATTERNS

Defining energy

We all recognise energy: in a two year old child; in a waterfall; or in the potential of a box of matches. It is more difficult to define. William Blake wrote 'energy is perpetual delight'. A British charity poster claims 'Energy is the power of the people'. A terse dictionary definition states that 'energy is the capacity of matter or radiation to do work'. Einstein concluded that 'energy is equal to mass multiplied by the velocity of light squared'. The definitions themselves indicate the variety and complexity of energy issues. We make the connection between energy, work, power and achievement at the personal level. We often fail to recognise that the whole nature of a society, its influence in world affairs, and its social, political and therefore spatial organisation, is largely dependent on its energy choices and the energy resources it can obtain.

Global reserves

Figures 1.1 and 1.2 indicate the global reserves of coal, oil and gas. These fossil fuels, together with uranium are the main **primary energy** materials at the global scale. **Secondary energy** is generated from primary sources to create a form of energy suitable for consumption. Thus peat, oil, gas, coal, wood or waste materials can be used directly as fuel (primary energy) or burned in power stations to make electricity (secondary energy). Since there is no transformation, hydro-electricity and nuclear power are both forms of primary energy.

Coal is more widely distributed at the global scale, and estimates of reserves are quite reliable. Figure 1.3 on page 6 shows the production of uranium which is the fuel that produces nuclear electricity. These reserves have not been fully estimated.

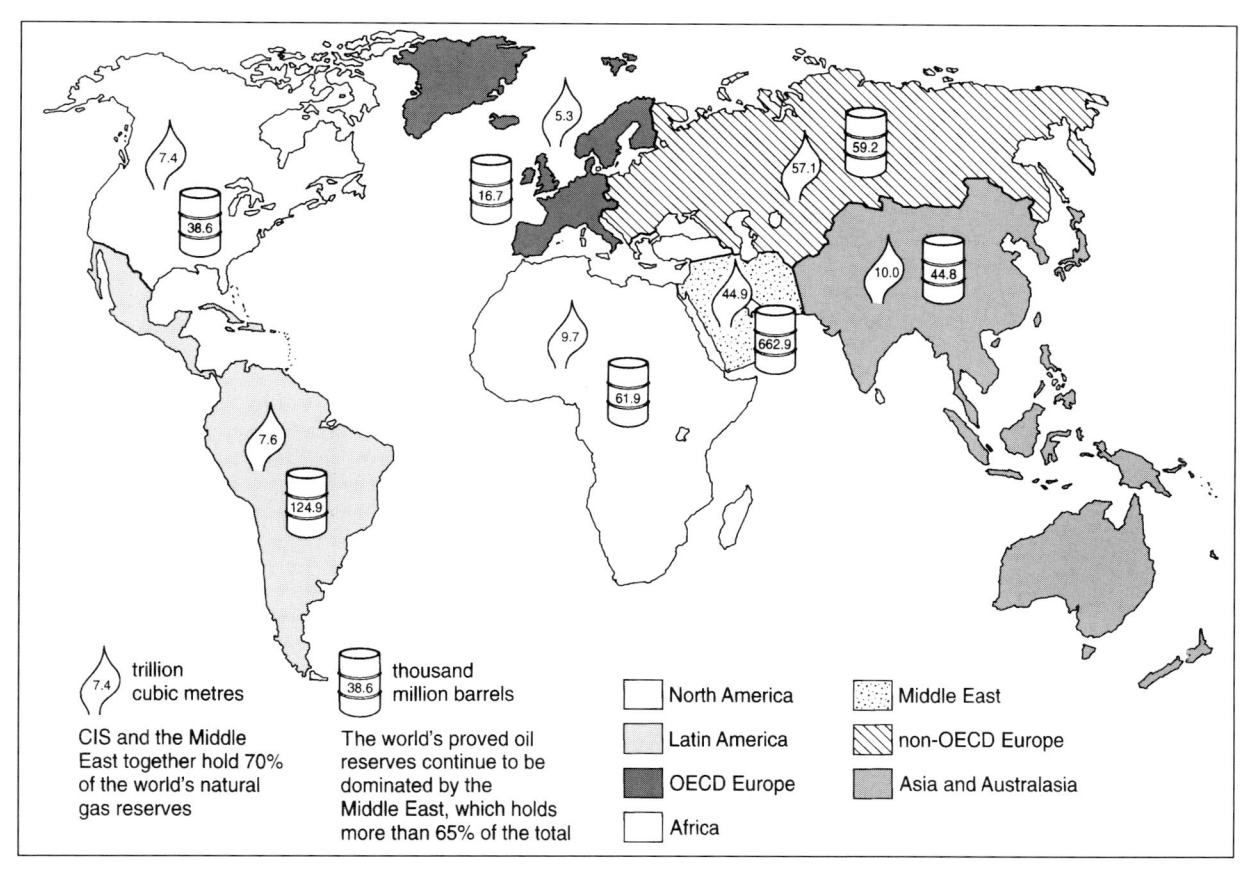

Figure 1.1 *Proved reserves of oil and gas by global region, 1993*

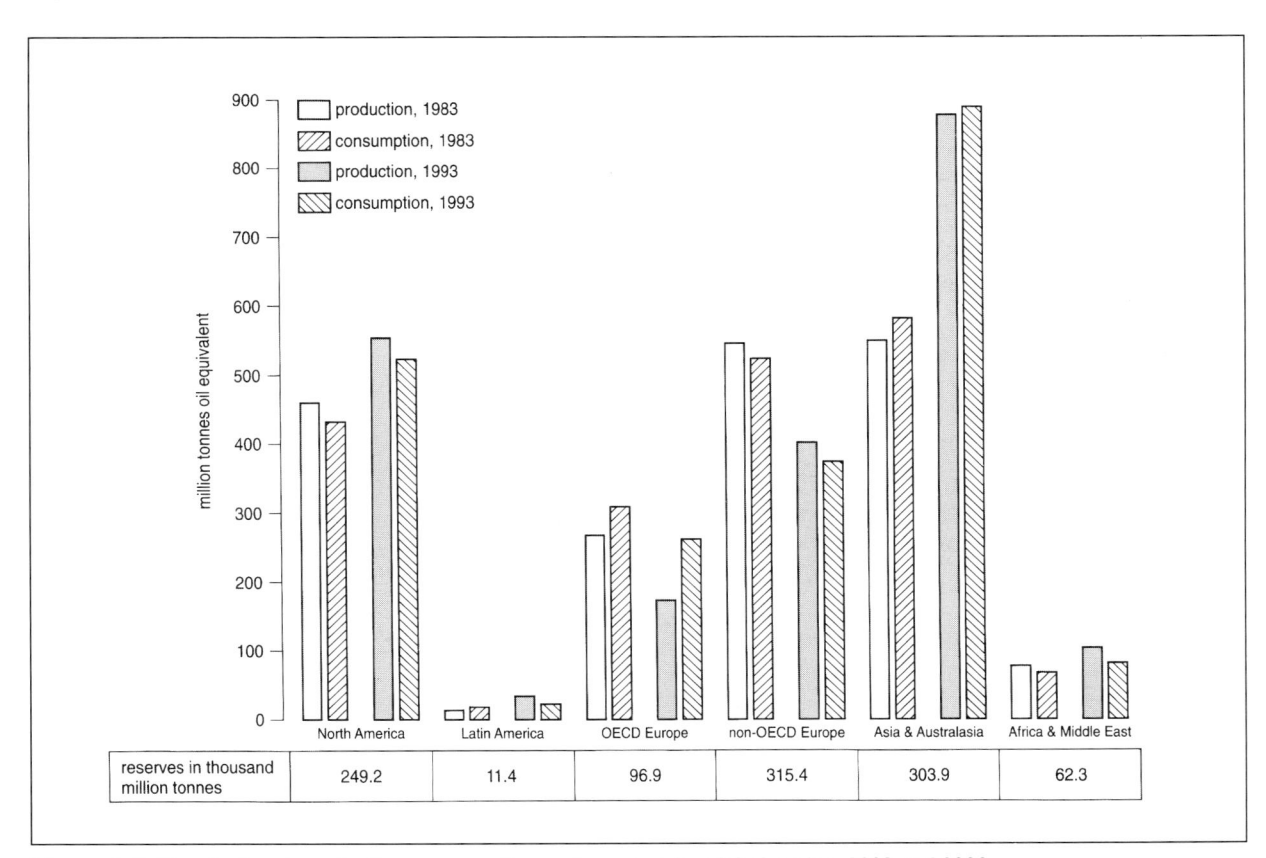

Figure 1.2 *Graph of coal production, consumption and reserves by global region, 1983 and 1993*

THE NATURE OF ENERGY 5

Figure 1.3 *Top ten uranium-producing countries, 1992*

COUNTRY	PRODUCTION 1992 (tonnes)	WORLD SHARE %
Canada	9385	26.4
Niger	2965	8.3
Kazakhstan/ Kyrgystan	2800	7.9
Uzbekistan/ Tadjikstan	2700	7.6
CIS	2600	7.3
Australia	2346	6.6
France	2127	6.0
USA	1808	5.1
South Africa	1769	5.0
Namibia	1692	4.8

> **Q1.** Why is most coal and gas consumed in the region in which it is produced?
>
> **Q2.** Why is information on reserves of uranium difficult to obtain?
>
> **Q3.** Why is the commodity of uranium much less significant in terms of quantity than coal or oil, in global trading patterns?
>
> **Q4.** Almost 30 per cent of the world's uranium comes from two companies; Cameco in Canada; and Cogema in France. Seventy per cent comes from ten companies. Why is the uranium industry such a concentrated energy industry?

Reserves of material energy only indicate potential; they must be exploited if a country is to benefit from them. Production and consumption statistics are easily obtained for the **commercial fuels**: coal; oil; gas; HEP; and nuclear electricity. Measurements of energy generated from **alternative renewable sources** such as wind, wave and solar sources, and from **biomass** sources (e.g. firewood) are almost impossible to collate and compare. It is difficult to quantify the contrasts in energy in a country such as Nepal which has 92 per cent of the population involved in low-technology agriculture, and highly mechanised societies where most wealth is generated by cognitive skills. Comparing the commercial statistics is only part of the picture.

The correlation between a society's wealth and its consumption patterns is easily observed. The higher the GNP, the more there is to spend on energy. Of course, energy resources create wealth in the first place – consider the emergence of the Middle Eastern countries this century from arid areas of limited significance to oil-rich global powers. Wealthier societies not only consume more energy, but also tend to draw on a greater variety of energy sources – Nepal, which relies on firewood for 98 per cent of its energy needs, can be contrasted with any of the countries listed in Figure 1.5.

Figure 1.4 *Production and consumption of oil for selected countries, 1993*

COUNTRY	PRODUCTION	CONSUMPTION
	(measured in thousand barrels daily)	
United Kingdom	2085	1790
Russian Federation	6995	3585
USA and Canada	10745	16410
Japan	+	5455
Indonesia	1530	790
Nigeria	1910	NA
China	2900	2965
Venezuela	2565	440
Papua New Guinea	125	+
Saudi Arabia	8695	NA
Total Middle East	19280	3655
India	560	1295

NA: Not available   + Statistically insignificant

The gap between production and consumption shown in Figure 1.4 indicates one of the most significant global economic patterns. Fuels are commodities which are not evenly distributed, and they can be traded. Coal, gas and uranium are not traded as much as oil. It could be argued that our whole world economic order is based on the trading relationships which have resulted from the concentration of oil in some countries.

> **Q5.** Use statistics from an atlas or the Geographical Digest to see if there is a correlation between GNP per capita and oil consumption. Use either Spearman's Rank Correlation Coefficient or present the information on a graph.
>
> **Q6.** Why do you think that Papua New Guinea does not consume any significant quantity of oil when it started producing it in 1991?
>
> **Q7.** Developed societies do not rely on biomass energy sources to any great extent. Why do you think this is?
>
> **Q8.** Is there a correlation between a country's wealth and the quantity and diversity of energy consumed? (Use an atlas to obtain GNP figures.)

Figure 1.5 *Commercial primary energy consumption*

ENERGY SOURCE (million tonnes of oil equivalent and % of country's total energy use)

COUNTRY	OIL	NATURAL GAS	COAL	NUCLEAR ENERGY	HEP	TOTAL
USA	787.5 39.5%	523.8 26.0%	494.7 25.0%	164.2 8.2%	25.7 1.3%	1996.0
CIS	178.6 24.7%	358.2 49.5%	140.8 19.5%	30.7 4.2%	15.0 2.1%	723.3
Venezuela	19.2 42.6%	22.0 48.8%	0.3 0.6%	–	3.6 8.0%	45.0
China	143.3 20.1%	14.2 2.1%	541.5 76.1%	0.7 0.1%	12.4 1.7%	712.1
Japan	252.6 55.4%	50.7 11.1%	79.3 17.4%	64.6 14.2%	8.7 1.9%	455.8
Philippines	14.1 89.2%	–	1.4 8.9%	–	0.3 1.9%	15.8
United Kingdom	84.1 39.1%	60.5 28.1%	50.7 23.6%	19.3 9.0%	0.4 0.2%	215.0
France	92.0 39.4%	29.2 12.5%	14.0 6.0%	93.4 40.0%	4.9 2.1%	233.5
Australia	32.0 37.1%	15.5 18.1%	37.2 43.1%	–	1.5 1.7%	86.3
Brazil	63.2 65.2%	4.0 4.2%	10.0 10.3%	0.1 0.1%	19.6 20.2%	96.9

ISSUES OF ENERGY

On a world scale: debt, dependency and global warming

In 1973 the Organisation of Petroleum Exporting Countries (OPEC) quadrupled the price of oil. This was the Arab response to the external support given to Israel in the Arab-Israeli war. It would be simplistic to say that this 'oil shock' precipitated a new economic order, but it was a significant factor in changing the global economy. International monetary relationships were such that the less developed nations had started borrowing cheap money on the Euro-dollar market. This was at a time when their commodities were commanding high prices.

The **petro-dollars** the Arab states had earned from their increased oil profits were deposited in the European banks. In order not to lose money on interest paid to the oil sheikhs, the banks had to recycle the money. The pattern of cheap loans to the Economically Less Developed Countries (ELDCs), facilitated by the system set up by the World Bank, was adopted. Thus money was loaned to ELDCs, firstly to pay for the increased price of imported oil and later the loans were absorbed by arms purchases (this was at the height of the arms race) and prestige development projects. This borrowed money did not generate profits in the ELDCs. Indeed much of it was held abroad by wealthy individuals. The classic example of this was the Marcos regime in the Philippines. Whilst oil did not cause the debt crisis, it intensified a pattern of borrowing which was to widen the gap between rich and poor countries.

A second OPEC shock followed the revolution in Iran. This doubled the price of oil in 1979. Demand in the industrial world fell. ELDCs saw the price of their commodities slump and with recession, interest rates soared. The banks which had made loans to these countries were faced with a dilemma. Unlike nations, or the World Bank, which had granted long-term loans, these banks could not cancel the debts, in a gesture of 'aid'. This was commercially unacceptable. A pattern developed whereby each country in debt reached a crisis, and admitted it could not pay. The International Monetary Fund (IMF) stepped in, and rescheduled the loans on condition that the country agreed to a series of austerity measures to help save money. Thus a relationship between a debt-ridden 'South' and a creditor 'North' evolved. This explains why Venezuela and Mexico are still in debt today although they have 6.3 and 5 per cent of global oil reserves respectively.

The oil shocks had other effects. Oil-dependent countries such as Japan started to diversify their energy sources. Exploration for oil in the North Sea and Nigeria intensified. Oil products, notably fertiliser, became expensive and pushed many developing countries further into debt. Today the inflation rate is related to the price of oil. Saudi Arabia has 25.9 per cent of the world's oil reserves, Iraq 9.9 per cent, and Kuwait, Iran and Abu Dhabi, all have over 9 per cent; these countries remain a focus of interest in global politics because of their oil wealth.

Energy relationships have affected the global economy and they are also connected to the global ecology. The **Greenhouse Effect** is thought to be a consequence of rising levels of carbon dioxide, methane, chlorofluorocarbons (CFCs), nitrous oxide and ozone in the lower troposphere. These molecules, together with water vapour, absorb infrared energy and heat from the surrounding air. Molecules of methane, largely a by-product of termite activity and rice production in the tropics, are 30 times more effective in this absorption process than carbon dioxide. Carbon dioxide is more abundant, however, contributing to 72 per cent of the warming with concentrations rising since the nineteenth century. The present acceleration rate is 0.5 per cent each year. One-third of all CO_2 emissions in the United Kingdom come from fossil fuel-burning power stations.

Predictions of actual temperature changes are difficult to make due to the two potential cooling influences of volcanic dust and sulphate aerosol. These may cause reflection of incoming sunlight and 'seed' clouds, making them whiter and thus causing further reflectivity. In 1988 the Intergovernmental Panel on Climate Change (IPCC) was jointly established by the World Meteorological Association and the United Nations Environmental Programme. Its brief was to assemble and assess information on the science, impacts and response options of global warming. The IPCC estimate is that there has been a global rise in temperature of 0.45 °C since 1900. Their report formed the basis of the agreement signed at the United Nations General Assembly (UNCED) in Rio de Janeiro in 1992.

Predictions related to global warming abound. Rising sea levels, monsoonal disturbance and increased cyclone and storm activity are just some of these and are all difficult to test and measure. Plants known to be sensitive to temperature and carbon dioxide levels could play a significant role in research and the economic motive could be important. For example, consider the implications of global warming for the forested areas of Scotland. Seventy per cent of all forested areas are planted with Sitka Spruce which takes 60 years to mature. Developed from British Columbia stock, the trees need 120 days below 5 °C to produce buds the following spring. Unless the trees can adapt, the economic consequences of rising air temperatures could be severe.

Until recently, the problem of global warming has been addressed by trying to limit CO_2 emissions. The scale of afforestation necessary to absorb enough CO_2 to counterbalance rising emissions makes it an unrealistic option. Deforestation at the global scale has, ironically, never been greater. Biotechnological solutions are the focus of research. Algae which digest CO_2 emissions from power stations have been developed in Japan and Britain but the systems are still at the experimental stage and are space-consuming and expensive.

> **Q9.** Find out the sources of CO_2 emissions. Why do you think that the Rio agreement did not set country by country targets for the reduction of CO_2 levels? What would be the impact of such targets on developing countries?
>
> **Q10.** Why was the United Nations Environmental Programme in 1987 more successful in setting targets for the reduction of CFCs?
>
> **Q11.** Why is the limiting of methane production a less acceptable way of addressing global warming?

The regional scale: acid rain, electro-magnetic fields and insects

It is difficult to separate the environmental issues related to energy from the social and the political, and it is arbitrary to separate the global from national or local scales. Pollution such as atmospheric and oceanic knows no national boundaries, and global energy problems can affect individuals at the local level. Despite the UN 'sulphur protocol' which agrees limits of sulphur emissions from power stations, it has been left to individual countries to cope with the effects of **acid rain**. This is acidification of rain caused by sulphur dioxide and nitrogen dioxide emitted from coal-burning power stations, road transport, and industrial processes. Norway, for example, spends £20–30 million each year on liming rivers and lakes in an attempt to neutralise water acidity. Acid rain is well-documented, but recent studies suggest that **acid flushes** are not being monitored by spot sampling and that these events, which are most damaging to river ecology, are not being measured.

Recent concern over the decline of the Scottish salmon catch, from 600 000 fish caught in 1968 to 166 000 in 1993, suggests acidity levels may be making conditions for spawning intolerable. Figure 1.6 illustrates this decline in the Dumfries region.

Electricity transmission can visually disturb landscapes. There is concern that there may be a connection between childhood cancers and proximity to the electro-magnetic fields associated with power lines. Whether this link is proved or not, American companies, fearing litigation, are avoiding residential areas when locating overhead or underground power lines. In Britain, **no-cost options** are being taken when siting new lines.

There has been much debate in the *British Medical Journal* over the suggested link between leukaemias and cancers in the Sellafield area, where the nuclear reprocessing plant, formerly called Windscale, is sited. The scientific illustrator, Cornelia Hesse-Honegger has drawn insects found at sites near Chernobyl, Windscale, Three Mile Island, and nuclear power stations in her native Switzerland. Her observations of mutated, deformed insects in many of these locations brought howls of protest from the scientific establishment. As scientists counter each other's arguments with more science, is it being left to other observers, painters and poets, to express unease at the issues of nuclear power? Hesse-Honegger may be proved wrong, but her observations must now be given a fair scientific test before being dismissed.

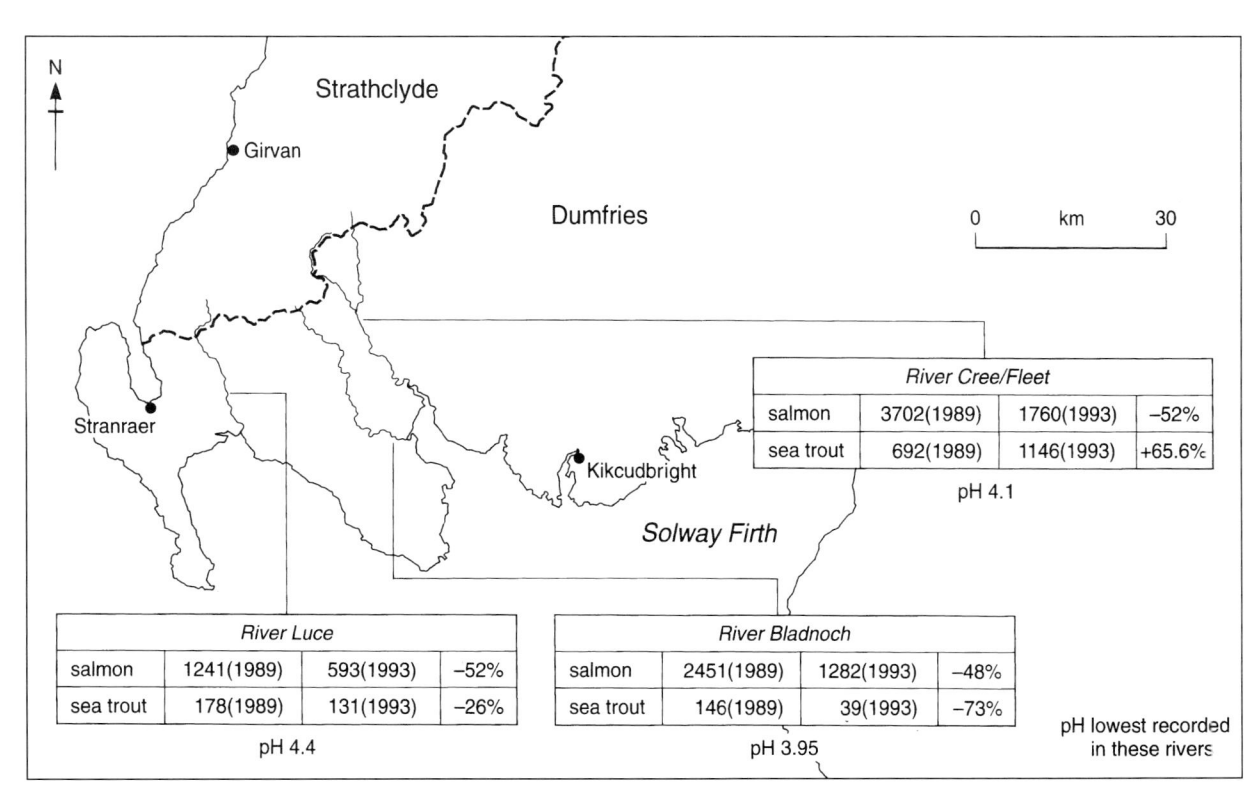

River Cree/Fleet			
salmon	3702(1989)	1760(1993)	–52%
sea trout	692(1989)	1146(1993)	+65.6%

pH 4.1

River Luce			
salmon	1241(1989)	593(1993)	–52%
sea trout	178(1989)	131(1993)	–26%

pH 4.4

River Bladnoch			
salmon	2451(1989)	1282(1993)	–48%
sea trout	146(1989)	39(1993)	–73%

pH 3.95

pH lowest recorded in these rivers

Figure 1.6 *Rivers and their acid levels in Dumfries, Scotland*

Figure 1.7 *Decommissioning this coal-fired power station at Gravesend, Kent, is cheap and simple, unlike the closure of nuclear power stations*

The local scale: work and worries

All energy decisions affect people at the local level. Power lines, spoil heaps, gas pipes, power stations, mines, and fuel handling facilities all bring **negative externalities** to residential areas. Most of these features are in place for at least one generation, but, like all industries, they change and decline. The **decommissioning** of nuclear power stations, by encasing them in concrete, is likely to have a negative impact on the landscape and security implications well into the future.

Energy uses also reflect the way a society is organised. In Uganda, the sale of firewood is 'women's work', and the collecting of fuel for each household absorbs up to 5 per cent of the working time of women and children in most of East Africa. Contrast this with Britain, where the energy industry is one of the largest employers in the country, highly centralised and male-dominated. Increasingly, people in the West are becoming aware that the way they use energy can affect the global ecology and that individual decisions are related to larger economic systems. A poster with the words 'Think global, act local', widely circulated in Los Angeles during 1988, stressed this link.

Q12. *Refer to a fieldwork techniques book and define 'spot sampling'. How could this technique be used to monitor acid flushes?*

Q13. *Why are 'acid flushes' the most likely way in which sulphur dioxide is introduced to the rivers in the north of England and in Scotland? (Consider the prevailing wind direction.)*

Q14. *Why are no-cost options a popular choice for power lines in Britain?*

Figure 1.8 *Sizewell B nuclear power station. A society's energy choices are reflected in the human landscape*

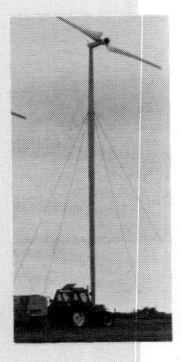

SECTION 2

Releasing energy

KEY IDEAS

- **To utilise each form of energy, a specific form of technology is required.**
- **All material forms of energy: fossil fuels, biomass, and nuclear fuels, produce waste which can damage the environment.**
- **Renewable energy can be generated from various natural and, by implication, sustainable sources. These forms of energy generation, whilst producing little pollution, have their disadvantages.**
- **Systems for generating renewable energy are at different stages of development, from the mature, commercially viable wind systems to the less developed photo-voltaic systems.**

FOSSIL FUELS

Coal

Coal, a combustible sedimentary rock consisting of plant remains, is one of the oldest non-renewable fuels. Spatially associated with sediments ranging from the Devonian to the Tertiary, a general rule is that the older the coal deposits, the higher the heat and lower the ash content. Tectonic movements, however, affect this relationship, and also determine the type of mining that can be undertaken.

Open-pit or open-cast mining involves removal of the overburden (surface earth), its subsequent dumping on adjacent land and systematic removal of the coal. Strip mining is one form of this. Production and recovery rates are high, costs per tonne are low, but large areas of land may be affected and need to be restored when mining is completed. Augering, a form of drilling used where the seams are inaccessible, is expensive. Underground mining, the oldest commercial means of production, has become increasingly **capital-intensive**.

Coal-scuttle Britain?

Perhaps because of its long history, and its significance in powering the Industrial Revolution, coal has had a central place in Britain's economy. The coalfields were associated with urbanisation, industrialisation and social change. A new capital-owning class, who had obtained mining rights, emerged. This industrialisation was not without its cost: in Oldham, in the 1850s, one miner in five died from accidents at the coal face. Problems in the coal

Mining (Scotland) Limited

1 Westfield Link
2 Lambhill
3 Longannet
4 Blindwells Extension
5 Rosslynlee
6 Damside
7 Dalquhandy
8 Airdsgreen
9 Piper Hill
10 Chalmerston

RJB Mining

11 Colliersdean
12 Stobswood
13 Ellington
14 Linton Lane
15 Plenmeller
16 Ryehill
17 Broughton Lodge
18 N Selby
19 Stillingfleet
20 Riccall
21 Wistow
22 Whitemoor
23 Gascoigne Wood
24 Prince of Wales
25 Kellingley
26 Long Row
27 Rockingham
28 Rossington
29 Maltby
30 Harworth
31 Arkright Colliery Reclamation
33 Point of Ayr
33 Welbeck
34 Rainge
35 Thoresby
36 Clipstone
37 Bilsthorpe
38 Calverton
39 Smotherfly
40 Kirk Revised
41 Nadins
42 High Cross
43 Bleak House
44 Asfordby
45 Daw Mill

Coal Investments PLC

46 Markham Main
47 Annesley/Bentinck
48 Silverdale
49 Hem Heath
50 Coventry
51 Cwmgwilli

Independents

61 Monktonhall Mineworkers Limited
62 Hatfield Coal Company
63 Betws Anthracite Limited
64 Goitre Tower Anthracite Limited

Celtic Energy Limited

52 East Pit Extension
53 Gilfach Iago
54 Ffos Las
55 Nant Helen
56 Helid Colliery
57 Kays & Kears
58 Great White Tip
59 Derlwyn
60 Llanilid West

■ colliery
● operating open-cast sites

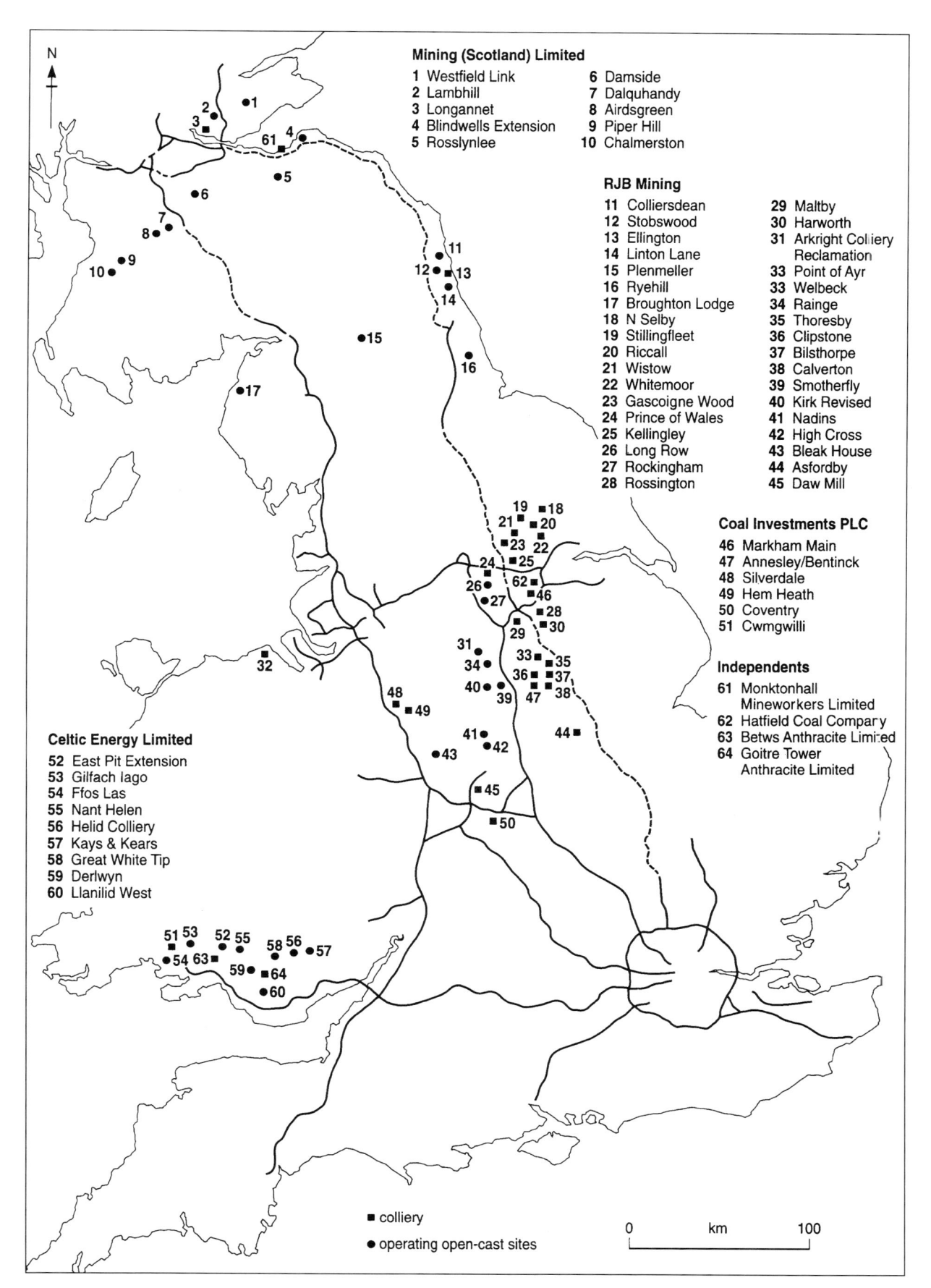

0 km 100

Figure 2.1 *British Coal: operational deep mines and open-cast sites in 1994*

industry developed after peak production in 1913, and were highlighted by the 1939–45 war. When the new Labour Government nationalised the industry in 1947, forming the National Coal Board (NCB), 800 businesses were amalgamated to form the largest employer in the country.

Despite strong demand for energy in Europe after the war, Britain failed to expand coal mining. The shortage of skilled labour, the cost of capital investment, and the continued working of pits which were not economically viable contributed to a near doubling of the cost of coal in the NCB's first decade. These problems paved the way for the Government's plan to use nuclear power, and the shift from coal-burning to oil-burning power stations. Nevertheless, the 'Plan for Coal' in 1956 attempted to expand coal output and surplus coal accumulated. There was a search for new markets but competition was fierce; the new European mines were still operating more efficiently, and had better management-labour relations. In addition, America had secured a place in the European market during the post-war energy-deficient years.

To add further to the declining demand for coal, its two main consumers, the electricity and iron and steel industries, both developed greater fuel-burning efficiency. Railways phased out steam, and natural gas came on stream in the late 1960s and so the business of coal gasification plants was lost. The domestic market dropped. After the London smogs, which precipitated the **Clean Air Act** in 1956, coal had an unhealthy image. The 1960s saw much urban clearance, and the new buildings were more suited to the use of other fuels. The British Government helped an ailing industry by direct financial help. A £415 million capital debt was written off, and the Government agreed to pay a large proportion of the social costs of colliery closures (Coal Industry Acts 1965 and 1967). Other measures to protect the industry were the banning of coal imports, a duty on fuel oil and the stipulation that the Central Electricity Generating Board (CEGB) burnt more coal, with the Government subsidising the extra cost. Disputes over wages dogged the industry, which was unable to meet the challenge of the oil shock, despite the NCB's plan for expansion.

The Thatcherite political climate of the 1980s emphasised the free play of market forces, and highlighted the very problem the industry faced, that of securing markets. New mines like Selby threatened over-production. When a pruning of the industry began in 1984, the miners' strike started in South Yorkshire. The combined factors of surplus coal stocks; spare oil-fired electricity generating capacity; the lack of support from the profitable Nottinghamshire mining area; and the willingness of the Government to support the National Coal Board, meant the strike lasted almost a year and defeated the National Union of Mineworkers (NUM).

With power in the hands of the renamed 'British Coal' management, mine closures continued. Productivity increased, assisted by sophisticated computer-controlled technology. In 1993, the Whitemoor mine in the Selby complex produced a record-breaking 1.8 million tonnes of coal. This mine closed later the same year. The workforce shrank from 181 000 in 1984 to 44 215 employees in 1992/93. The geography of coal today is shown in Figure 2.1 on page 13.

The Coal Industry Act of April 1993 provided the legislation for the privatisation of the industry. Critics claim it could lead to a regression to pre-nationalisation conditions. British Coal was privatised at the start of 1995 and the private mines report to a new 'Coal Authority'. Fragmentation of the industry leading to competition between pits when there is an increased threat from imported low sulphur coal may prove unhelpful. Whilst coal may not compete with gas in the mid-1990s, it may once again become a significant energy source if gas supplies decline.

After underground seams have been extracted coal mining can bring slag heaps, dust, heavy road and rail traffic, noise pollution, severe landscape disturbance and subsidence. Nevertheless the 'community values' of mining areas were recognised, or feared by successive governments. One of the themes of twentieth-century British writing, from D H Lawrence's *The Rainbow* to Dennis Potter's screenplay, *The Singing Detective*, is the migration of the bright youngster from a mining background to a different strata of society, a different landscape. The mining community, real or imagined, contributes much to the national sense of place, and perhaps this explains why coal has had such an emotive place in political debate.

Q15. *Outline all the factors which have contributed to a decline of coal production in Britain this century.*

Peat

Peat is vegetable matter accumulated at temperatures of less than 5 °C, the threshold for coal formation. Peat deposits are less compressed than coal and found near the surface. When first cut there is a high moisture content, but when dried the peat can be used as fuel with 30–60 per cent moisture. Indonesia has the world's largest reserves, with 200 billion tonnes, and the highest producer is the **CIS**, with an annual output of 17.5 million tonnes. It is unlikely that the current use of 7.5 million tonnes per annum in Ireland, where a number of power stations burn peat, will be sustained much beyond the turn of this century.

Oil and gas

Petroleum is the generic term for hydrocarbons derived from organic material and deposited during sedimentary rock-basin formation. These range from methane gas, through liquid oils to the solid petroleum waxes.

Oil is not spatially associated with coal. Any society that has changed its energy resource base from coal to oil, such as the United Kingdom, will therefore show a shift in energy patterns. If Europe led the coal-fired Industrial Revolution, the USA pioneered the new social and economic order of an oil-based, high-energy society. Commercial drilling for oil began in the USA in Pennsylvania, in 1859. As Figure 2.2 indicates, supplies were plentiful and tapped early.

> **Q16.** *Translate the data from Figure 2.2 onto a line graph and then describe and explain the trends shown.*
>
> **Q17.** *At what point did the USA lose its world lead in oil production? To what would you attribute this shift in leading producers?*
>
> **Q18.** *Why has production dropped in the former USSR?*
>
> **Q19.** *During which decade did the Middle East become world leader in oil production?*

To keep prices strong, consumption, particularly through car ownership, was encouraged. Oil is a highly mobile fuel; it can be piped, shipped by tanker or moved by road or rail. In turn, its use brought about high mobility in an affluent society. As Figure 1.4 (page 6) shows, the USA remains a high oil-consuming society.

The North Sea bubble

In 1959, a 2600 m deep test drill discovered methane near Groningen in the Netherlands. This led to exploration of the North Sea bed in the 1960s. Since it is less than 200 m deep, the North Sea is classified as a continental shelf, and under International Law belongs to the countries adjoining it. Exact boundaries were agreed by the eight interested countries, and the first strike of gas in British waters was in 1965. Oil was first landed offshore in Norway in 1974 and in Britain in 1975. In 1979, at the then price level, almost £15 million gross value of oil was being drilled each day.

So why is this story of successful exploration and drilling referred to by some as a failure? There are several reasons for this. Firstly, only Shetland and Orkney managed to turn oil into long-term developments. For example, in 1971 the Shetland County Council Act gave the council the right to charge taxes on oil shipments and developments, giving a resultant income of £50 million per year for 20 000 citizens. No such legal act protected Aberdeen. It is true that by 1979, 56 000 jobs had been created in this city and that Grampian region recorded high levels of employment and above average incomes for the 1970s and 1980s. However, the oil city saw a huge rise in house prices, and the cost of services was inflated by the influx of wealthy oil workers from outside the area, for only 40 per cent of the North Sea labour was drawn from the UK. When the oil industry retracted, the property and wages 'bubble' burst. In attempting to sustain investment in the city, Aberdeen tried to become the 'Houston of Europe', developing services such as the University's Department of Offshore Medicine which is now a private enterprise.

The 'bubble' did not appear to benefit the rest of Scotland. Throughout the 1970s, the Scottish National Party sourly observed the continued decline of

Figure 2.2 *World oil production in million tonnes*

YEAR	FORMER USSR AND EASTERN EUROPE	NORTH AMERICA	LATIN AMERICA	MIDDLE EAST	AFRICA	ASIA	WORLD TOTAL
1890	3.5	3.5	–	–	–	–	9.9
1910	12.1	29.2	0.2	–	–	2.8	44.5
1930	23.6	128.6	26.0	6.4	0.3	7.8	192.9
1950	42.1	284.3	88.2	87.8	2.3	13.6	519.3
1970	372.3	563.3	237.0	706.0	302.5	89.6	2286.5
1985	617.1	578.4	334.0	532.5	252.6	158.5	2789.5
1993	406.1	504.1	406.5	944.7	330.7	329.3	3164.8

Figure 2.3 *The decline in fishing and its shift to Peterhead meant that the dock areas of Aberdeen absorbed the new oil-related shipping with little change*

traditional industry (Figure 2.4). The profits, it was argued, were slipping south to fund tax-cuts and cushion the country's debts rather than being used to invest in new industry. In Wales the new terminals for imported oil at Milford Haven similarly did not generate industry in the region.

Figure 2.4 *Scottish National Party sticker from 1978*

Secondly, North Sea oil supplies have their limits, which will be exhausted in the near future. By 1991 46 offshore and 19 onshore fields were in production, netting 91.3 million tonnes of oil. Oil profits generated 1.5 per cent of the United Kingdom's GNP in 1991. Income to the treasury from taxes and royalties amounted to £100 million in 1991/92. Despite the issuing of new licences for exploration, it is largely accepted that North Sea oil has peaked, and that the energy map will not change significantly, unless Atlantic exploration is successful.

Thirdly there are environmental problems with oil: spills can be damaging to marine and estuarine life. The social cost of the North Sea venture, with the pioneering of offshore technology, has been high. Lives were lost, or destroyed by stress, in the drive for profits. The Piper Alpha accident, reflected in the 1986 figures in Figure 2.6 on page 18, killed 300 men and the subsequent Cullen Inquiry exposed inadequate safety procedures.

Natural gas reserves

Unlike oil, gas can be found near coal, dissolved in oil (**associated gas**), or trapped in a layer above an oil reserve. Associated gas was, until recently, flared off. It can be used to maintain pressure in an oil well, by being pumped back into the oil reservoir. Its

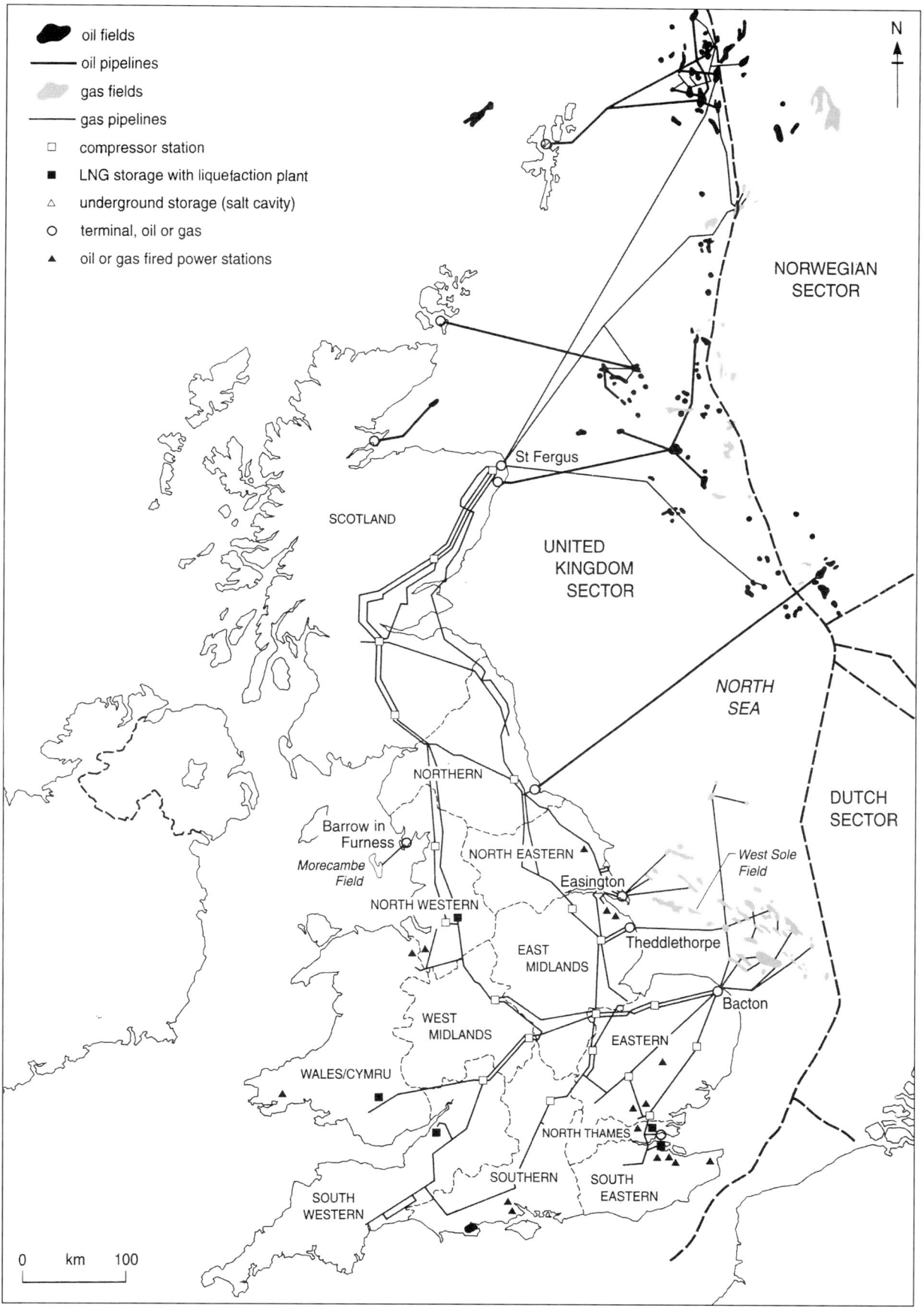

Figure 2.5 *Oil and gas in the UK*

Figure 2.6 *Oil spills reported to the Department of Energy (UK), 1982–1991*

YEAR	1983	1984	1985	1986	1987	1988	1989	1990	1991
NO OF SPILLS	62	47	87	166	254	259	291	345	234
ANNUAL TOTAL TONNES	186	130	310	3540	516	2627	511	899	192

retrieval is fraught with technical difficulties. Dissolved gas has a variable oil/gas ratio and must be removed before the oil can be safely transported.

Gas can be moved in the form of Liquefied Petroleum Gas (LPG) by refrigerated tanker ships. Similarly gases such as propane and butane, which liquefy under pressure at normal temperatures, can be moved by tanker. This is a hazardous enterprise; in 1984 in Mexico City an explosion from an LPG distribution centre killed 500 people. Pipeline systems are expensive to lay and maintain, and only feasible if a country is certain of its markets or has large reserves. Discovery of the Continental Shelf reserves has changed the British gas industry into a distributor of primary fuel. The gas industry, nationalised in 1948, was organised into a series of 'Gas Boards', each serving its own separate area. Gas works tended to be located in the major urban centres and a national grid did not emerge until the 1960s with the importing of cheap Liquid Natural Gas (LNG) from Algeria in 1964. Seven of the 12 area boards were linked to supply this gas.

Subsequent gas discoveries in the West Sole field off Humberside stimulated a new distribution system, centred at new terminals at Bacton, Theddlethorpe and Easington. St Fergus was added later, and the Morecambe field was tapped via a terminal at Barrow-in-Furness in 1985. As the southern supplies have declined, the northern Frigg field has become more significant in the North Sea. The distribution system shown in Figure 2.5 on page 17 served 34 offshore fields and one onshore field, in 1991. Britain reflects global trends in the 'dash for gas' (see page 27) as new gas-fired power stations are built. This has long-term consequences for the coal industry, and may create dependence on foreign gas when home supplies decline in the future.

Q20. *What event in the early 1970s may have accelerated the pace of exploration for oil in the North Sea?*

Q21. *Why do you think that the oil industry did not create 'growth poles' in the UK in the way that the coalfields encouraged growth of new settlements and industries?*

Q22. *Contrast the figures for 1983 and 1991 in Figure 2.6. What is the underlying trend of the number of oil spills and quantity of oil lost?*

Q23. *Refer back to Figure 1.1 (page 5). What regions are likely to be significant in supplying gas in the future?*

NUCLEAR POWER

Nuclear power, a non-renewable form of energy, represents one of the most significant technological developments in the twentieth century.

Some very heavy atoms, like the **isotope** uranium 235 (U235), when bombarded with a neutron, can be 'split', a process causing a chain reaction of subsequent fissions and a huge release of energy. In a nuclear reactor this fission is controlled, and the heat generated used to produce steam, which in turn operates a turbine and generates electricity. Nuclear fuel has a unique energy/weight ratio; 0.5 kg of uranium has the same energy as 1000 tonnes of coal. Radioactivity from nuclear fuels, however, is potentially lethal to all forms of life and the fuels, waste products, and equipment used in handling them, must be sealed off from the **biosphere**. For this reason Denmark and New Zealand do not permit nuclear facilities.

The nuclear fuel cycle shown in Figure 2.7 outlines the complete process from mining to final disposal.

Uranium is found in a variety of conditions and deposits with just 1 per cent uranium ore are considered economically viable.

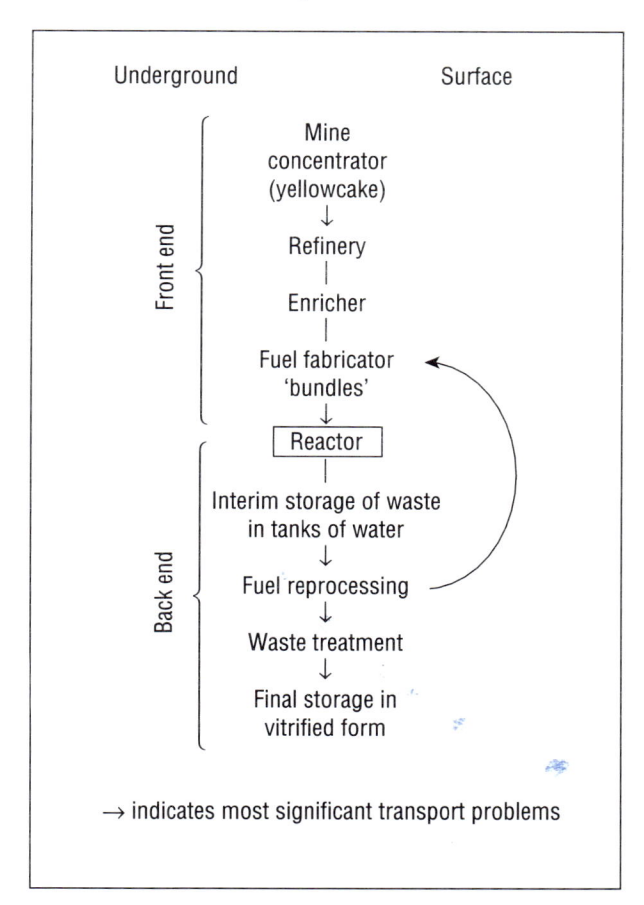

```
                Underground        Surface
           ┌                Mine
           │            concentrator
           │             (yellowcake)
 Front end │                  ↓
           │               Refinery
           │                  ↓
           │               Enricher
           │                  ↓
           │            Fuel fabricator
           └              'bundles'
                             ↓
                          Reactor  ←──┐
           ┌                 ↓        │
           │        Interim storage of waste
           │           in tanks of water
           │                 ↓
 Back end  │          Fuel reprocessing ──┘
           │                 ↓
           │           Waste treatment
           │                 ↓
           │          Final storage in
           └            vitrified form

    → indicates most significant transport problems
```

Figure 2.7 *The nuclear fuel cycle*

Uranium is usually concentrated at the site of the mine before being moved in the form of 'yellowcake' to the refinery. Refining removes impurities and enriching increases the proportion of U235. (The element uranium consists of three isotopes; U238, U235, and U234. Of these only U235 is fissionable, and less than 1 per cent of the uranium consists of this isotope.) Nuclear reactors never use pure uranium, but all except the CANDU design (pressurised heavy water reactor) require enriched fuel. Fuel fabrication prepares the fuel for placing in a reactor by encasing pellets of UO2 in tubes of zinc alloy or stainless steel and assembling them in 'bundles'.

Even at this 'front end' of the cycle there is a risk. A uranium tailings pond collapsed near Church Rock, New Mexico, on 16 July 1979, distributing radioactive thorium 230 through an 80 km stretch of the Puerco River.

The reactor (Figure 2.8) needs a **moderator** to control the speed of the neutrons, allowing fission to take place. Water, carbon in the form of graphite, and 'heavy water', (deuterium oxide found in ordinary water in a ratio of 1:6500) are used as moderators

Regulation of the process, as well as start-up and shut-down of the reactor is carried out by control rods. These are made of a material which will absorb neutrons, such as boron or cadmium. When the control rods are inserted into a reactor, they absorb a proportion of the neutrons being released, reducing the level of fission. To start the reaction, the control rods are gradually withdrawn. One of the problems of a nuclear reactor is, however, that even

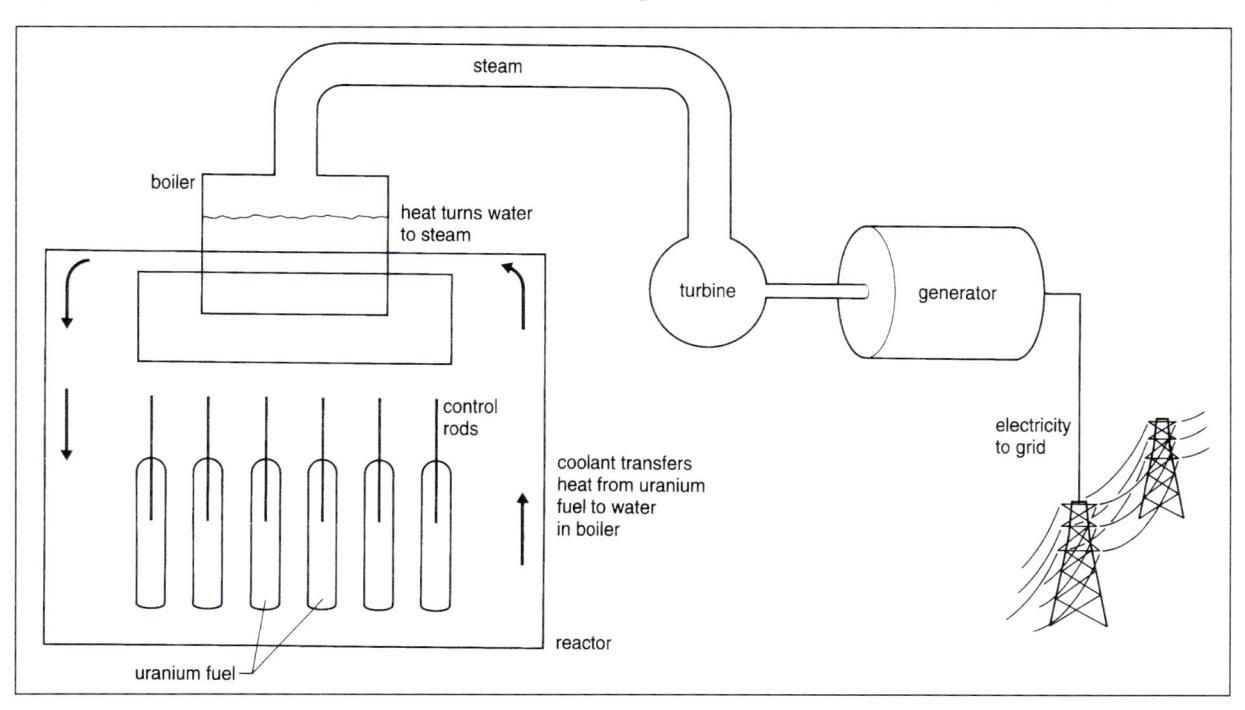

Figure 2.8 *Components of a nuclear reactor*

once the control rods have been inserted, a reactor continues to generate heat.

Heat produced is transferred to the boiler in the power station by the **coolant**. Gas-cooled reactors use carbon dioxide or helium, whereas liquid-cooled reactors use water, heavy water, liquid sodium or a molten alloy of sodium and potassium. The names of the different reactors relate to the type of coolant, the moderator material or the fuel type.

Plutonium is a product of all nuclear reactors which use uranium, and this plutonium can in turn contribute to the fission chain reaction. In a **fast breeder reactor** neutrons transform the uranium 238 present in the reactor into plutonium. Highly enriched fuel (up to 75 per cent uranium 235) is used in a very compact core, without a moderator. It is surrounded by uranium 238 which absorbs the neutrons and 'breeds' plutonium. This intense process can present cooling problems.

Plutonium and reusable U235 can be recovered from the spent fuel rods, in a reprocessing plant; an expensive process carried out by remote control. If the rods are not reprocessed they are immersed in water in special containers to cool for several decades. This high level waste will then be enclosed in glass (vitrified), and buried at a suitable site free of **groundwater**, faulting and **geothermal** heat.

> **Q24.** Why is nuclear power more suited to providing **base load** than **peak demand** electricity?
>
> **Q25.** Why are many nuclear power stations located by coasts or large rivers?
>
> **Q26.** In what ways can nuclear electricity be considered renewable, although uranium is in finite supply?
>
> **Q27.** Why have proposals for long-term storage of high level nuclear waste at the surface been rejected?

Classification of nuclear waste is most often on the basis of the level of radioactivity. Low level waste is 'dispersed' in the environment within agreed limits. The cumulative effect of these low-level radionuclide particles is not known. Environmental groups point to the disappearance of seven species of bird from the Ravenglass estuary near the Sellafield reprocessing plant. Intermediate-level waste is buried. Drigg, near Sellafield accepts this waste. There is still debate over suitable sites for the long-term disposal of high-level waste. The shipping of nuclear waste to ELDCs keen to earn foreign currency, by offering storage facilities, is a potential hazard.

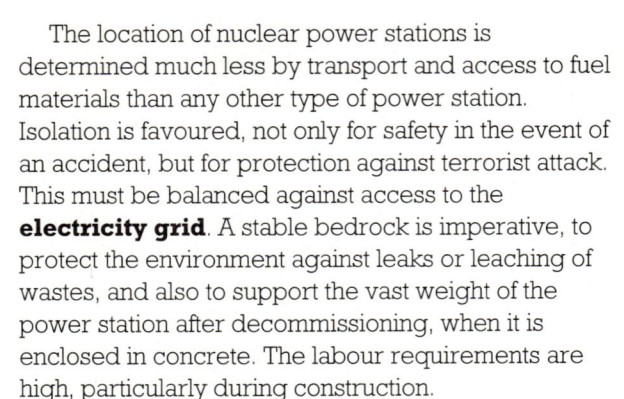

The location of nuclear power stations is determined much less by transport and access to fuel materials than any other type of power station. Isolation is favoured, not only for safety in the event of an accident, but for protection against terrorist attack. This must be balanced against access to the **electricity grid**. A stable bedrock is imperative, to protect the environment against leaks or leaching of wastes, and also to support the vast weight of the power station after decommissioning, when it is enclosed in concrete. The labour requirements are high, particularly during construction.

The nuclear fuel cycle, from mining to disposal of waste, is subject to much greater security and bound by more laws than any other energy use. Indeed critics of nuclear power are often concerned not only with safety, health and the environment but with the constraints which nuclear power, with its high levels of security, necessitates. It is also argued that the sophisticated knowledge required to construct and administer a nuclear power station places too much responsibility in the hands of very few scientists. Aldous Huxley warned of this connection between nuclear energy and style of government in his foreword to *Brave New World* predicting **totalitarianism** in those countries that had harnessed atomic power.

Chernobyl: a postscript

On 23 April 1986 reactor four of the Chernobyl nuclear power station in the Ukraine exploded, sending tonnes of nuclear fuel and graphite into the lower atmosphere. One and a half million Ukranians were exposed to radiation, which then swept north east in a cloud over Belorussia, and across to Sweden. Here, the high levels of radiation measured forced Moscow to admit to the accident. The town of Pripyat next to the station was not evacuated until 36 hours after the accident, although the town's telephone wires had been cut the morning after the fire.

It will take many years to judge the full impact of the accident. In the five years following the accident no new reactors were commissioned in the former USSR. The event probably helped speed up the disintegration of the Union, for after the event, many of the conservative Ukranians judged Moscow in a harsher light.

An unsafe RBMK design (a light water graphite moderated reactor), the lack of procedures for dealing with reactor failure, the lack of hazard awareness, and the poor communications in the industry were blamed for the accident.

In 1991 the Ukranian Government voted to close the remaining three reactors by 1995, but ever-decreasing supplies of coal, oil and gas from Russia, on which it is dependent, make this unlikely. Believing the RBMK design to be unsafe, in July 1994 the G7 countries offered the Ukraine £130 million to close the power station. The Ukranians refused the offer. Earlier that year Russia had resumed its exports of fuel rods to the plant after a blockade to force the Ukranians to place its activities under International Atomic Agency safeguards.

The environmental voice in the CIS is becoming quieter as economic pressures on the new states grow, but accidents continue. In April 1992 a container of radioactive deposits exploded, contaminating an area of 120 km^2 around the secret Tomsk plant in Russia. Yet in the same year Russia announced the intention of increasing its nuclear capacity from 22 000 to 37 000 megawatts (mw).

About 200 000 Ukranians were evacuated from their homes and the Ukraine Government spends 12 per cent of its annual income on coping with the effects of the Chernobyl accident. None of this goes towards decontamination. It is difficult to quantify the full effects of the accident on the nation because the then Minister for Health was forbidden to cite radiation as a cause of death. This ban was only lifted in 1991. The Research Institute of Haematology in Minsk observed a six-fold increase in thyroid cancers between 1986 and 1991. Few hospitals have computers to assist in the enumeration of the effects of the accident.

The cloud of radioactive dust was washed down by heavy rain in the parts of Britain shown in Figure 2.9. The areas which received most radioactivity were where the rain fell heaviest, six days after the explosion. Caesium 137, with a half-life of 30 years, is the most dangerous isotope. Peaty uplands are vulnerable because the soil contains few minerals which bind to caesium. Five years after the accident, 410 farms in Wales and 144 farms in Cumbria were still affected. Sheep have had to be removed from the upper pastures and grazed in less contaminated valleys to allow the material to be flushed from their

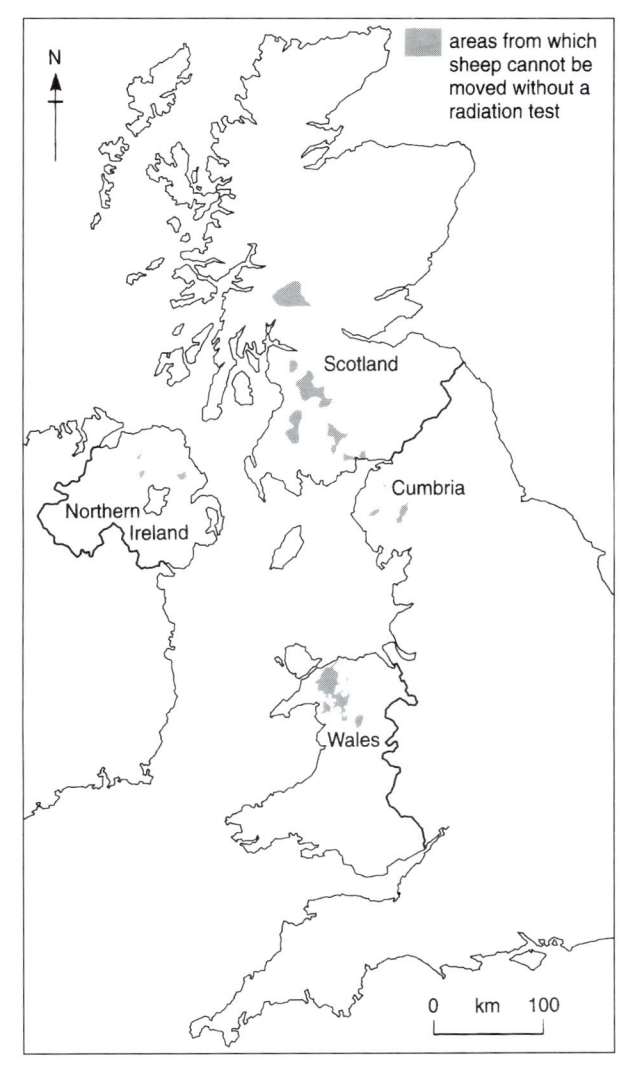

Figure 2.9 *Areas of the UK from which sheep cannot be moved without a radiation test, 1991*

bodies. What is an inconvenience and a loss of income to farmers thousands of kilometres away must be having effects as yet unmeasured nearer Chernobyl. 'We could become a nation of mutants' comments a leading Ukranian writer. With the splintering of the Soviet Union it is too easy for this concern to be dismissed as nationalist hysteria.

RENEWABLE ENERGY

Renewable energy supplies are defined as those which harness energy in the natural and, by implication, sustainable environment. Impact on the environment is minimal, compared with the use of nuclear or fossil fuels. Since concern for rising population levels and increasing consumption of resources was voiced at the Earth Summit in Rio de Janeiro in June 1992, renewable energy has had more attention. Technology to tap most of the sources is now available, although it is not always commercially viable and is often limited to specific locations.

RELEASING ENERGY **21**

Hydro-electric power

HEP is a well-tried form of renewable energy. Electricity is produced from generators driven by hydraulic turbines. To give the required 'head' and constancy of flow, dams are built. Multi-purpose schemes use the reservoirs created for recreation, water supply or irrigation. Once dams and turbines are in place, HEP produces the lowest cost electricity available. Such schemes are often seen as the key to the economic development of a region.

Over time, however, silting by rivers reduces reservoir capacity. The initial expense, and the construction time of the HEP scheme are also problems. Other negative effects can include disruption to navigation, lowered soil fertility downstream, and the economic and social effects of displacing people from the area to be flooded. Often the product, electricity, is irrelevant to the lives of those cleared from the land, and benefits only industrial or urban areas. The criticisms of the Three Gorges Scheme in China and the Narmada Scheme in India focus on this and on the huge scale of the projects. Countries where HEP is most appropriate are those with a suitable terrain, and a level of economic development that allows even distribution of the benefits. Examples include New Zealand and Norway.

Small-scale schemes are those with a capacity of up to 5 mw. They have low running costs, high reliability and long life. Now that 'crossflow turbines' allow 'run of river' schemes with a low head, there is a wide availability of sites, and these schemes are being actively promoted in the UK.

Tidal power

The principle of tidal energy is the trapping of water in an estuary, at high tide. When the sea water outside the barrier has dropped to low tide the impounded water in the estuary is let out through low head axial turbines, generating electricity. A high tidal range is essential for this technology; the Bay of Fundy in Canada and the Severn Estuary are two of the best tidal sites in the world. Figure 2.10 shows the location of some potential barrage schemes in the UK where the mean tidal range is 3 m or more.

If the Severn barrage was built it could supply between 6–8 per cent of the annual demand for electricity in England and Wales. A £4.2 million development study of the potential barrage was completed in 1990. The capital cost, the study estimated, would be as high as £10 billion, and the construction time at least six years. The scheme

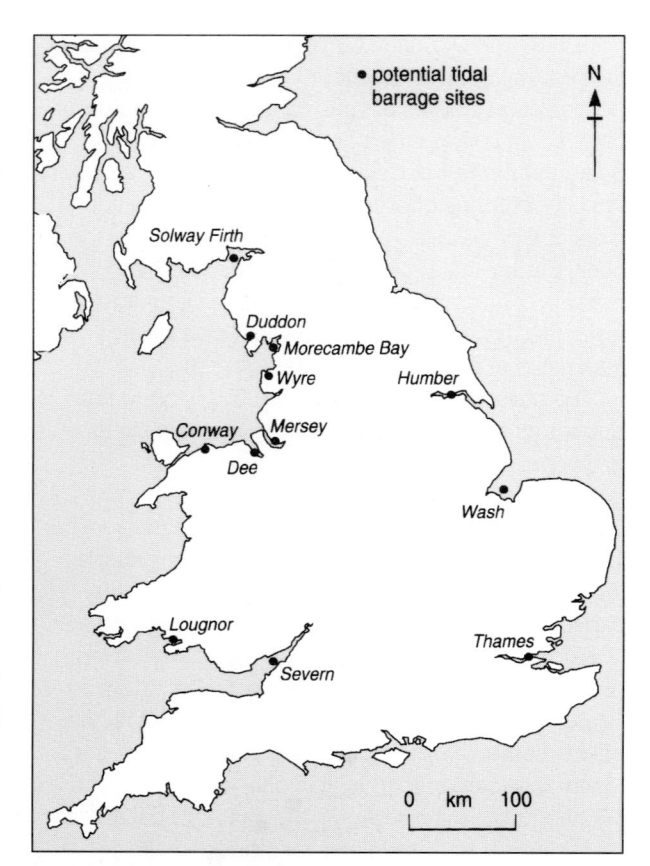

Figure 2.10 *Suitable tidal power sites in the UK*

would have a life of 120 years, however, and could be very cost-effective.

The saving to the environment in terms of reducing fossil fuels must, however, be balanced against disturbance to local estuarine ecosystems. Britain's location means that the estuaries are very important feeding grounds for migratory birds. The hydrodynamics of the estuary would also be affected, changing the dispersal of effluents. The turbidity of the water could be reduced, so increasing **primary biological production**. An increase in phytoplankton in the water could affect the whole food chain. What is bad for 'nature' ultimately affects humans. It is as difficult to quantify the negative environmental effects as it is to estimate the positive economic gains such a scheme might have.

Solar radiation

Although solar radiation is free its collection is difficult and expensive. Making solar energy commercially viable is also frustrating; the laws of thermodynamics and photosynthesis are understood but technical problems abound. Solar energy can only be successful in areas with reliable year-round sunshine and minimal weather interference. A high proportion of homes in Greece, Israel and Australia use solar water heaters. These **active solar devices** are expensive to produce, however, and their manufacture consumes high quantities of energy!

Passive solar devices are more applicable in northern Europe, and since approximately 45 per cent of Europe's energy is used in buildings, they should make substantial savings. Principles include siting buildings so that the main glazed areas face within 30 degrees either side of south, reducing glazed areas on the north side, and supplementing the design with energy-efficient features such as insulation and sensitive heating systems. Casual energy gains from machinery, lighting and cooling are thus maximised, and contribute to the heating of the building. The University of East Anglia has completed new student residences which were designed using these ideas and also use solar heating.

Watches and calculators are powered by photo-voltaic (Pv) solar cells which change light into electricity with no moving parts. It is hoped that this technology can be extended. The greatest potential for large-scale generation lies where there is high solar energy and a high demand for power. California powers most of its air-conditioning with this technology. Germany's 'thousand roofs' programme funded by federal and local subsidies, will provide individual customers with up to two-thirds of the cost of supplying such systems. In Switzerland all schools can apply for a grant to install these Pv panels.

Other methods of photo-conversion using photo-biological, photo-chemical and photo-electrochemical techniques are being explored, with most research focusing on the creation of stable systems.

Geothermal power

Heat from the earth is accessible only in a few countries. Iceland supplies most houses with thermally heated water, which is also used in domestic heating systems. Steam is also used to generate electricity in Iceland and New Zealand. Geothermal aquifers are tapped in France and Hungary for district heating, and in Italy for generating electricity.

In granite batholiths there are high increases in temperature with depth. The Hot Dry Rock project in Cornwall attempted to generate electricity from

Figure 2.11 *This 100 kw photo-voltaic grid connected installation was put into operation in Switzerland at the end of 1989. The 830 m long band of photo-voltaic modules annually produces 115 000 kw/h. This is the world's first example of the combination of sound barriers and photo-voltaic grid connection*

steam obtained by injecting water to depths where it was heated and piping it to the surface. Following an economic feasibility study, and with pressure on the Government to support alternative energy forms that were already commercially viable, Government funds for this project have been withdrawn, as have contributions to the EU Hot Dry Rock research project at Soultz-Sous Forêts near Strasbourg, where a pilot generating facility is envisaged.

Biofuels/biomass

Energy can be produced from crops and plants, and from animal wastes. This form of harnessing energy is traditionally associated with societies with a low level of economic development where there is widespread direct use of these materials such as firewood. Other less direct uses are also available.

In Brazil, sugar cane is converted into fuel as a substitute for petrol. The Government was strongly criticised for its choice of plantation-grown sugar cane with an ethanol yield of 3500 litres per ha per year, rather than cassava which can be grown by individual farmers. Cassava, however, has an ethanol yield which is one-third less than sugar cane and requires an additional saccharification process before being fermented.

The crops require energy to harvest, transport and process, and only those factories fuelled by the crop itself make an energy gain. With sugar cane, for example, the crushed material left after juice extraction, bagasse, can be used to fuel the factories producing alcohol. Other suitable crops include sweet sorghum, sweet potatoes and corn, all having an ethanol yield of at least 2000 litres per ha per year.

Fuels derived from these crops produce no net increase in emitted CO_2, and have become a focus of interest in biomass-to-energy schemes. In the UK biomass accounts for 4 per cent of all primary energy. Livestock wastes are highly polluting, and **anaerobic digestion** (AD) can help reduce nitrate leaching, odour and river pollution caused by slurry applied as manure. Combustion of these wastes can also produce heat and power, and an ash rich in phosphates and potash. New environmental considerations have made farmers reassess the energy options of these fuels and co-operative schemes could encourage economies of scale.

Q28. *Why do you think some politicians in Brazil favoured the use of cassava?*

Q29. *Consult an atlas to identify areas where crops with the highest energy potential are grown. What difficulties might these countries encounter in tapping this energy source?*

Wind energy

Like the waterwheel, the windmill had its place in medieval European society. Large mills could produce up to 30 kw of energy, enough for a small community. At the end of the nineteenth century, Britain had 10 000 windmills in operation. Small-scale schemes have been used throughout this century in the USA and Australia in isolated rural areas to power domestic lighting, radios and water pumps. They are also appropriate for island communities; the Greek island of Kythonis has a small park of five 20 kw wind turbines complemented by a 500 kw diesel power station and a photo-voltaic plant.

Large wind turbines, producing wind for a grid system, are usually of the horizontal axis design with three or four blades. About 300 kw of electricity can be produced by a machine with a 25 m blade. Wind must reform behind the machine; in a wind farm the closest machine is seven to ten diameters away from the next one. Wind speed is important, for the power available in wind varies as a cube of its velocity.

The best sites are exposed, free of obstructions and subject to strong winds. Criticism focuses on noise, interference with radio and television reception, the scale of the enterprises, and the visual disturbance these 'farms' bring to the landscape. In the UK the Department of Energy has stated that 20 per cent of the country's electricity could be generated by wind by the year 2025.

Energy from waste

Some waste products in land-fill sites produce a natural gas, methane, which can be collected, piped and burned to produce useful energy. Sewage treatment plants can be powered by this methane by-product. Burning waste directly, however, can produce more energy. When burnt in incinerators designed to harness heat, garbage can produce a third as much heat, weight for weight, as coal.

The US National Energy strategy, published in 1991, called for a seven-fold increase in electricity from municipal waste by 2010. In the Netherlands there are 12 municipal waste plants, five of which generate electricity. In Britain, at the turn of the century, there were 70 incinerators generating

Figure 2.12 *Farmers gain rent from the operators, yet agriculture can continue uninterrupted by wind turbines*

electricity from municipal waste; these were closed when Local Authorities were withdrawn from the task of supplying power.

Incinerators have been criticised by environmentalists who argue that they discourage recycling and composting. However, they save on land-fill sites where 90 per cent of Britain's household waste is dumped. The danger of the possible production of dioxins is prevented by passing gases quickly through the 200–400 °C temperature band, where these gases form. Pollutants are absorbed with charcoal and lime injections, and the waste gases are filtered.

Wave energy

It has proved difficult to design devices to tap this extremely large source of energy. Such devices would need to be able to withstand storms and also transport the electricity generated to the shore. In Norway and the Isle of Islay land-based power stations have been designed. Waves enter at the bottom of a 20 m high cylinder; the column of water rises and pushes air through an air turbine at the top; the turbine is driven as air is sucked back down the tube when the wave passes. Such an oscillating water-column device produces 500 kw at Tostallen, north of Bergen in Norway.

Q30. *What would be the difficulties in imposing a 'polluter pays' policy for nuclear waste?*

Q31. *What are the site requirements for:*
   a) nuclear power stations;
   b) the generation of hydro-electricity; and
   c) wind farms?
   Why could the siting of all these types of power generation be subject to criticism on environmental grounds?

Q32. *Attempt to summarise the advantages and the design problems of each type of renewable energy.*

Q33. *Attempt a classification of the different forms of electric power generation according to:*
   a) wastes produced; and
   b) economic and social benefits to the area near the power station.

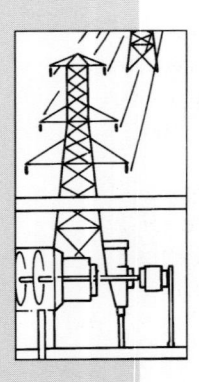

SECTION 3

Harnessing energy: the UK and Japan

KEY IDEAS

- Some societies have an explicit energy policy. Others, through political and economic policies, are less direct but just as influential in shaping their energy patterns.
- All energy patterns result in social and spatial patterns, observable at different scales.
- Many countries have energy problems. The nature of these vary with the country's history and level of development.

ENERGY POLICY IN THE UK: GREY OR GREEN?

The sensitive relationship between the economy, society and energy can be observed in the UK. The pattern has been one of shifting focus: from coal up to 1945, to imported oil in the 1960s, to the offshore fields and associated terminals in the 1970s and 1980s. The status of nuclear electricity has advanced and receded parallel to other changes: the oil shocks; the discovery of oil and gas; and the miner's strike. The story is one of stimulus and response. The capital-intensive nature of the nuclear power industry means that the response has often been slow. Nevertheless, it is possible to trace different spatial trends for each energy era. The distribution systems in the energy industry have changed. Coal, gas and, to a certain extent, electricity were strictly regional until the mid-1960s. National grids now exist for gas and electricity, while a pipeline system ensures wide distribution of oil.

Even though coal and gas were nationalised after the 1939–45 war, there was no co-ordination of these industries or any attempt to formulate an overall strategy. Nor has there been since. Energy policy has remained unstated, although protection of the coal industry was a significant policy after the 1967 Coal Act. Energy policy suggests intervention and control, and recent Conservative governments have not favoured this. Instead, free enterprise was the ethic in the 1980s and early 1990s. As in many areas of national life, the concern was for performance indicators, particularly those measured by financial cost. Privatisation, it was argued, increased competition, promoted efficient allocation of resources, and reduced government involvement in the economy. It is too soon to assess the spatial effects of privatisation, but it is worth reviewing the changes and particularly their impact on the electricity industry in England. In this way shifts in the geography of energy in the early part of the next century can be anticipated.

Privatisation of the electricity industry in England

The CEGB was a fully-integrated electricity utility until 1990, running power stations in line with its technical ability to match demand through each day. Thus nuclear and coal-fired power stations provided baseload, and surges in demand were supplied by peak power from gas and oil turbine units.

Privatisation in England in 1990 basically involved dividing the generating function of the CEGB between National Power, Power Gen, and Nuclear Power, and forming 12 regional distribution companies. The National Grid was made a separate company owned collectively by the distribution companies. National Grid has two pumped storage systems at Dinorwig and Tan-y-Grisiau to meet surges in demand. Its main role, however, is to control all transactions between generating concerns and distribution companies. It controls the high voltage transmission system, and it also purchases power according to a 'pool price' bidding system and the level of demand.

One consequence of privatisation has been the threat to the indigenous coal industry. British Coal had an agreement with the CEGB whereby 95 per cent of its coal requirements were purchased from British Coal. The CEGB was British Coal's biggest customer. However, in a desire to become more independent of each other, the new regional electricity companies have attempted to sell their expertise abroad and have sought out new generating capacity. The Combined Cycle Gas Turbines, quick and cheap to build, have been seen as the answer; they produce no sulphur dioxide (SO_2) and only half the CO_2 of equivalent coal-fired power stations. The switch to gas, which of course puts British Gas in a very powerful position, threatens to destroy the coal industry. It is difficult to envisage how competition will be structured in the industries that are 'natural monopolies' because of their **fixed track system** and capital-intensive natures. The electricity companies, like the railways, face perplexing times ahead as the arrangements to facilitate competition take effect. It could be argued that the energy industry is of necessity collaborative and it is interesting to note that no other European country has taken on this form of competition.

> **Q34.** *The increased costs of reprocessing used fuel at Sellafield have made Nuclear Power, once again, very expensive. What other reasons contributed to the original exclusion of Nuclear Power from the privatisation process?*

The dash for gas

Statistics from the Organisation for Economic Co-operation and Development (OECD) for 1991 suggest that the UK is one of Europe's worst contributors of sulphur oxides in the atmosphere. Coal-fired power stations contribute more than two-thirds of these emissions. Under the United Nations' Sulphur Protocol the UK has agreed to cut this pollutant to one-fifth of the 1980 level by the year 2010. To meet this obligation, Britain planned to spend about £2 billion on **retrofitting** flue gas desulphurisation plants on some of the largest coal-fired power stations, and low NOx burners to all 12 major coal-fired power stations.

Cheaper options have also been pursued. Part of the reduction has been achieved through importing low-sulphur coal, and by the building of ten gas-fired power stations. Only two coal-fired power stations have been fitted with desulphurisation plant: National Power's Drax in North Yorkshire and Powergen's Ratcliffe-on-Soar plant in Nottinghamshire, which together will burn 15 million tonnes of coal a year. There are hopes that the Government might permit the 'dirty' and part-time burning of coal elsewhere. On course to reach the 2010 target, the Secretary of State for the Environment signed a treaty in Oslo in 1994, reaffirming this commitment.

> **Q35.** *The emissions of sulphur oxides have been lowered in the UK, but what are the dangers of switching to gas?*
> **Q36.** *What factors determine the location of gas-fired power stations?*

The wind rush

In 1991, the British Government imposed a Non Fossil-fuel Obligation (NFFO) on the newly privatised electricity industry, whereby some electricity not generated by coal, oil or gas had to be purchased. Some argue that this is in order to encourage renewable energy; others maintain the measure is designed to safeguard an increasingly uncompetitive nuclear power industry. This policy is paid for by a tax which adds 0.2 per cent to electricity bills; 98 per cent of this money goes to nuclear power; only 2 per cent to renewable energy. Even this small proportion has encouraged applications for wind turbines. An added attraction is that applicants may gain grants from the **EC THERMIE Programme** to help establish their wind farms; the sale of electricity could thus be highly profitable. By May 1994 there were 400 wind turbines in the UK at 21 sites, producing 0.2 per cent of the country's electricity. Application for a further 230 sites had been lodged, the largest at Kielder in Northumberland for 200 turbines. Plans for these projects have been halted, for there has been a strong negative response to what Simon Fairlie, writing in *The Ecologist*, has called these 'White Satanic Mills'.

The issue has provoked controversy and two camps have emerged. The response of local communities has been, on the whole, to oppose the wind turbines and Figure 3.1 shows the main action groups in the UK for and against the wind turbines and their construction.

Wind turbines in other parts of Europe are used to encourage local self-sufficiency. In Denmark, 3500 turbines are owned by co-operatives, whose shares are restricted to local people and related to individual consumption. This can be contrasted with Britain where the wind farm idea has not been connected to local aspirations but has been placed in the hands of the corporations.

> **Q37.** Why do you think there are environmental groups on both sides of the wind farms argument?
> **Q38.** Which parts of Britain are most suitable for wind farms?
> **Q39.** Do you think the wind debate is only about preserving a landscape heritage or are other factors involved?

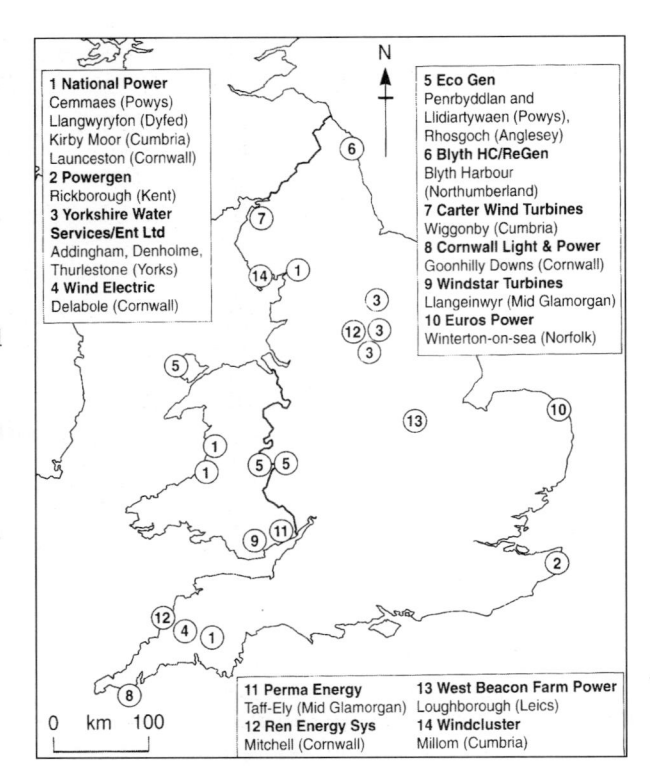

Figure 3.2 *NFFO contracted wind projects*

Rubbish issues?

It is not only in wind-swept areas that the NFFO has had an impact. Small-scale energy projects, encouraged by this obligation, will have an impact in various localities. In Deptford in East London the subsidy earns a new generating plant two and a half times the market price for electricity sold. This plant is fuelled by municipal waste. The subsidy expires in 1998, but by then the plant will be selling heat to neighbouring homes. The issues that this plant raises are important, and relate to the Royal Commission on Environmental Pollution recommending that the British Government should give targets to disposal authorities, to recover energy from municipal waste.

Figure 3.1 *The wind farm debate*

OPPOSING	PROPOSING
The Rambler's Association	Royal Institute of International Affairs
Council for the Protection of Rural England	Greenpeace
Council for the Protection of Rural Wales	Friends of the Earth
The Country Guardians*	The Green Party
Local communities and local environmental groups	British Wind Energy Association[+] lobbying on behalf of corporate investors who sell electricity
	Landowners

* led by Bernard Ingam, a consultant to British Nuclear Fuels, and George Ratcliff, former chairman of the Energy Research Liaison Committee
[+] this includes British Aerospace, Taylor Woodrow and Tomen International (Japan)

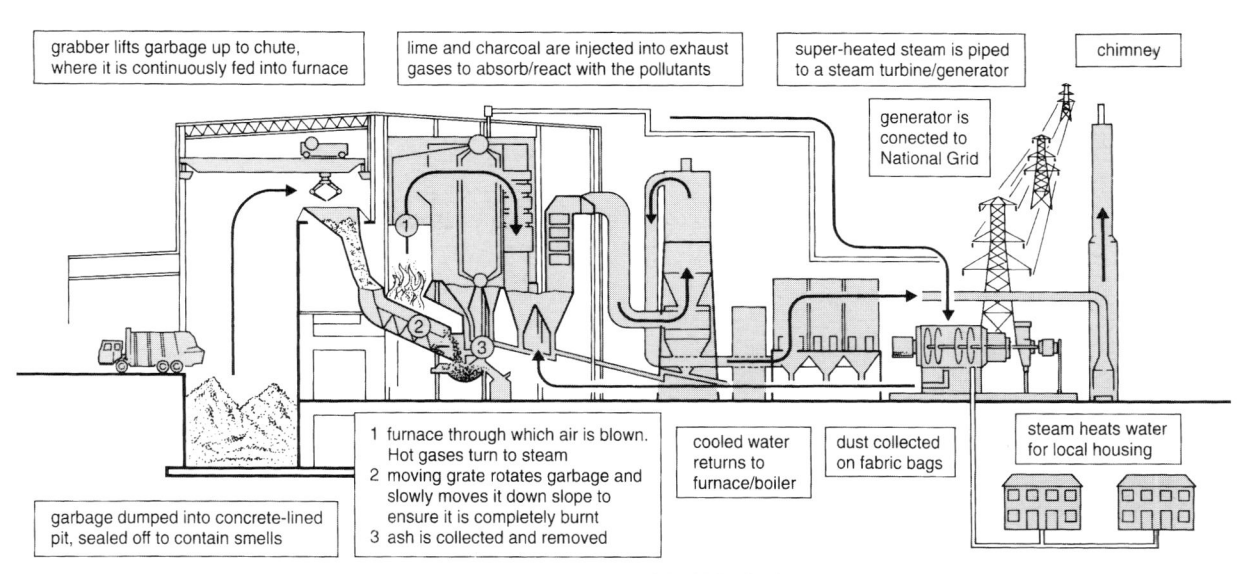

Figure 3.3 *The South East London Combined Heat and Power (SELCHP) Project*

Q40. Study Figure 3.3 and refer back to 'Energy from waste' in Section 2 (page 24). Outline the attitudes different groups might have towards such a project. Consider the opinions of, for example: the Local Authority; the residents near access roads to the plant; the local waste recycling group; and the power station shareholders.

Q41. Although these schemes save on land-fill space, why are some environmentalists opposed to the rubbish-to-energy concept?

Q42. These waste-to-energy plants tend to be located near or in urban areas, close to the material energy source. What other locational factors are important in the siting of waste-to-energy plants?

Q43. Study Figure 3.4 which shows the total energy flows for the UK in 1992 and attempt to define the main locations of the inputs and the outputs.

Q44. Attempt to predict what changes the following might have on the pattern shown: a) privatisation; b) VAT on domestic fuel consumption; and c) the environmental lobby.

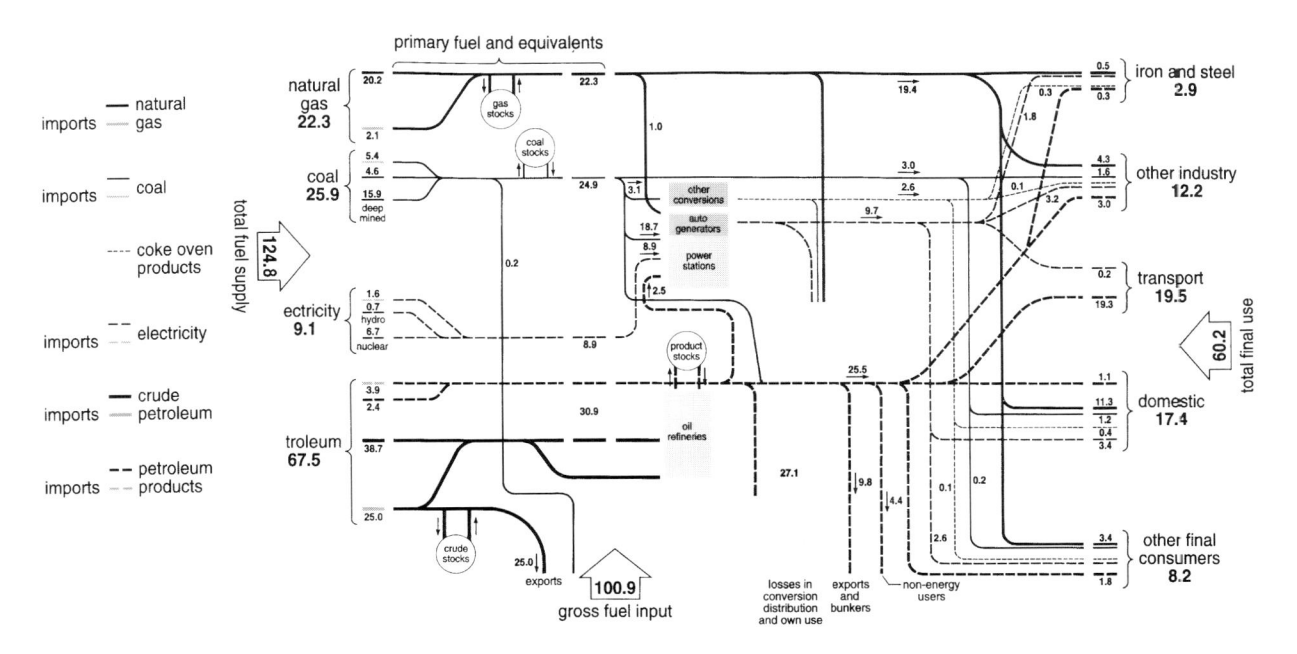

Figure 3.4 *Energy flows for the United Kingdom, 1992*

JAPAN: CATCH UP CAPITALISM

Japan has few energy reserves and its recovery and development since 1945 is one of this century's best examples of determination coupled with the effective use of aid, granted to Japan after the 1939–45 war. In the second half of the 1960s its economy achieved a mean annual growth of 11.8 per cent. Consumer spending, capital investment in steel, petrochemicals and other manufacturing soared. Energy demand increased at an annual average rate of 11.7 per cent between 1965 and 1973.

Japan's economic activity makes it a significant energy consumer at the global scale. World-wide, 17.4 per cent of oil, 28.4 per cent of coal and 73.6 per cent of Liquefied Natural Gas (LNG) is destined for Japan.

Japan was vulnerable to fluctuations in imported energy. The effect of the oil shocks, (see Section 1), can be seen in the levelling off of demand, as shown in Figure 3.5, and also in Japan's policy to diversify. In 1984, however, oil still constituted 59 per cent of Japan's primary energy sources.

The response to the two oil shocks was to conserve and promote efficient energy use. Sources other than the Middle East were sought to reduce dependency on one supply group. The shift from the Middle East is shown in Figure 3.6a. Nevertheless, the country still depends on imported energy and this dependence has grown as Figure 3.6b shows.

In order to eliminate short-term fluctuations, oil storage capacity was increased, and Japan can now hold 4 per cent of a year's supply. However, the increase in oil prices brought about recession in the more energy-intensive sectors of the economy. These industries, such as aluminium smelting and petrochemicals, declined and there was a parallel shift towards the 'soft' or 'knowledge-orientated' industries. High-technology industries such as microelectronics expanded. For example, between 1970 and 1980 the volume of integrated circuits produced in Japan increased on average by 50 per cent per year.

The Government moved towards a policy of diversification from oil, which was to be formalised in the 1980 'Law to promote Development and Application of Non-Oil Energy Alternatives'. This course has been difficult. The domestic coal industry had been in decline since 1950, precipitated by the poor quality of the coal and the difficulty in mining. Japan sought new, and more reliable markets. Today the USA and Australia supply most of Japan's coal.

In the diversification process, hydro-electricity was unlikely to contribute a greater proportion of primary energy, for most of the suitable sites had already been developed. The most recent HEP addition, the Naramata Dam on the Tone River north of Tokyo, is a multi-purpose scheme providing flood

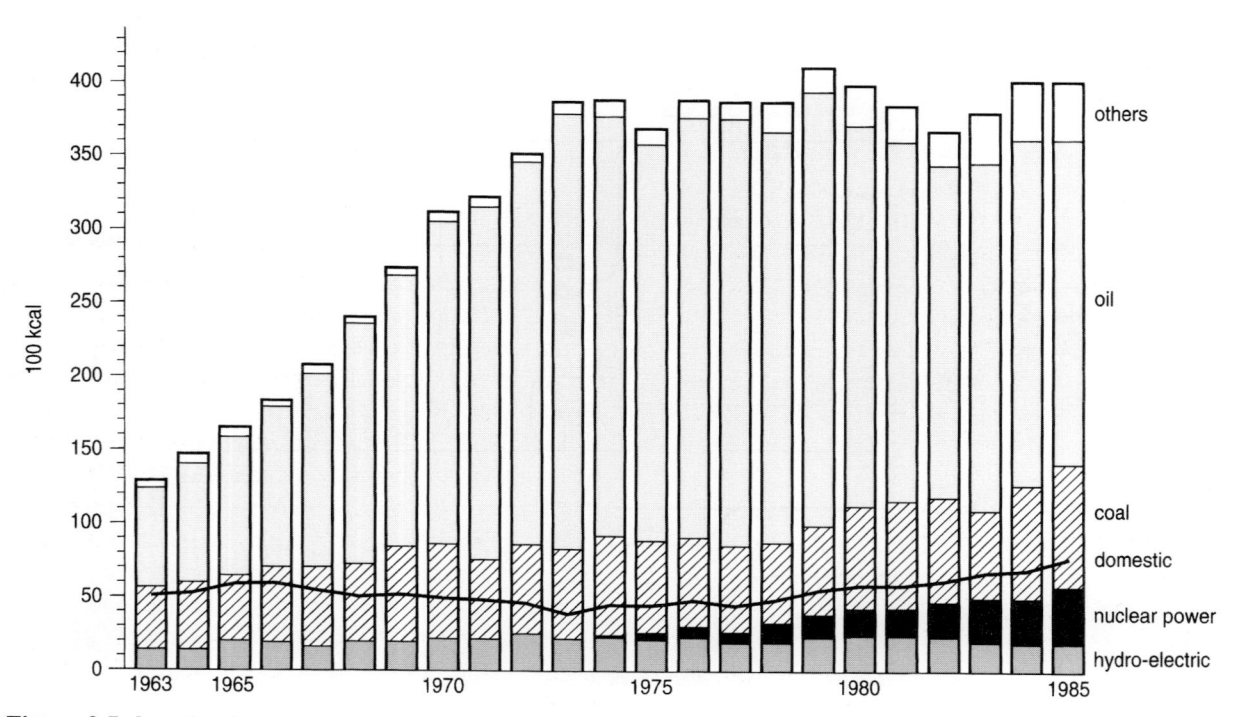

Figure 3.5 *Japan's primary energy supply*

30 HARNESSING ENERGY: THE UK AND JAPAN

a)

YEAR	1970	1980
	(per cent)	
Imported primary energy dependence	88.0	82.4
Primary energy dependence on oil	75.7	55.4
Dependence on oil imports	99.7	99.7
Imported oil dependence on Straits of Hormuz	75.1	66.3
Primary energy dependence on the Middle East	58.9	35.1

b)

(Fiscal years) (1 000 000 mill on kcal)

	1960	1970	1980	1989	1990
Coal	415.2	635.7	673.3	796.7	807.5
Petroleum	379.3	2298.9	2624.4	2674.3	2835.4
Hydro-electricity	157.8	178.9	204.8	211.2	205.1
Nuclear electricity	–	10.5	185.8	411.5	455.1
Natural gas, LNG	9.4	39.7	241.6	461.6	492.8
Total [1]	1008.1	3197.1	3971.7	4618.8	4861.6
Dependency on imports (pct)	43.4	84.4	89.8	92.0	92.6

Note: [1] Includes others

Figure 3.6 a and b Changing dependencies in Japan's energy supply

fuel cycle. Figure 1.5 (page 7) shows the significance of nuclear power in Japan.

In pursuing its nuclear programme, Japan has been beset with problems. As Figure 3.7 on page 32 illustrates, much of Japan is mountainous and there are only limited areas of flat coastal land where the bedrock is close to the surface. Proximity of nuclear processors to airline routes and military bases is prohibited for security reasons, and sites cannot be near areas of high population density. Areas prone to seismic instability, also shown on Figure 3.7, and tidal waves (**tsunamis** can reach 20–30 m high) are also unsuitable. In addition, the nuclear power industry has met with local resistance, particularly from fishers.

control and water for industry, agriculture and domestic consumption. The technology required to build this dam in an area of seismic instability made it one of the most expensive dams in the world (£600 million in 1990). Its contribution makes little change in the energy proportions.

With little scope for increasing oil and gas production capacity, apart from expanding LNG imports from South East Asia, Japan's options to diversify were limited. The choice was to increase nuclear, or renewable 'alternative' sources.

Japan, a 'hi-tech' society, has sought the technological solution, and has ruthlessly pursued expansion of nuclear power. In 1990 Japan was the world's fourth largest producer of nuclear power, following the USA, France and former USSR, and world-wide it has the fastest growing number of nuclear units. Despite this, the expansion aims of the 'Long Term Programme' drawn up by Japan's Atomic Energy Commission (AEC) are rarely fulfilled. Hokkaido Electric Power cut its prices to consumers in 1993, a reduction attributed to the new nuclear power units 1 and 2 at Tomari. Japan is now attempting to become self-sufficient in the nuclear

Japan's years of 'catch up capitalism' were associated with high levels of environmental degradation and pollution and it is anxious to improve its image. By its own admission, the country is the fourth largest emitter of CO_2 (by country). Typically, a high-technology solution is being attempted. The Research Institute of Innovative Technology for the Earth (RITE) was established in 1990, and this centre has produced a potential bio-technological solution to the problem of high levels of CO_2. A micro-organism, chlorococcum littorale, has been defined. This is a species of plankton capable of photosynthesising, and **sequestrating**, large quantities of CO_2.

In 1989 a White Paper on the Environment noted that levels of sulphur dioxide had decreased from 0.057 parts per million in 1970 to 0.011 parts per million in 1985, and levels of carbon monoxide from 6.0 parts per million to 2.4 parts per million in the same period. Levels of nitrogen dioxide had however slightly increased, and the standard of acceptance was relaxed from 0.02 ppm to 0.06 ppm in 1985.

As Japan addresses the problems of pollution from

traditional sources by stricter legislation, it is also supporting two projects. Firstly, the 'Sunshine Project' established in 1973 investigates alternative energy sources, in particular solar energy options. Currently though, only photo-voltaic technology has reached the mature, commercially-operable stage of development. Secondly, a parallel 'Moonshine Project' attempts to encourage energy conservation.

The problems of traditional energy sources and consumption patterns are being attended to. The concern now is that as the newer technologies, particularly nuclear power, are being created, new environmental issues are developing. It is too early to assess if the new technologies will leave an environmental legacy, as the fossil fuel technologies did in the 1960s and 1970s. The refusal of Hong Kong harbour to berth a Japanese ship carrying plutonium to Japan on 24 August 1993, for fears of spillage, may be a significant expression of disquiet.

Many Pacific Rim economies are now pursuing nuclear power, as Europe and the USA pause to reassess their programmes. Taiwan contracted its latest plant in early 1994, and Indonesia is investigating possibilities for a first site. The echo of the new economic term, 'Pacific Rim', with the old geographic term, 'Pacific Ring of Fire', is surely thought-provoking. Care must be exercised in the push for power in an unstable seismic environment. Already, at Marong in the Philippines, a nuclear power station lies unused. It was discovered prior to its opening that it had been built on a fault line.

Q45. *How has Japan attempted to become more self-sufficient in energy?*

Q46. *What are the implications of the geology for the siting of nuclear power stations in Japan and the disposal of high level nuclear waste?*

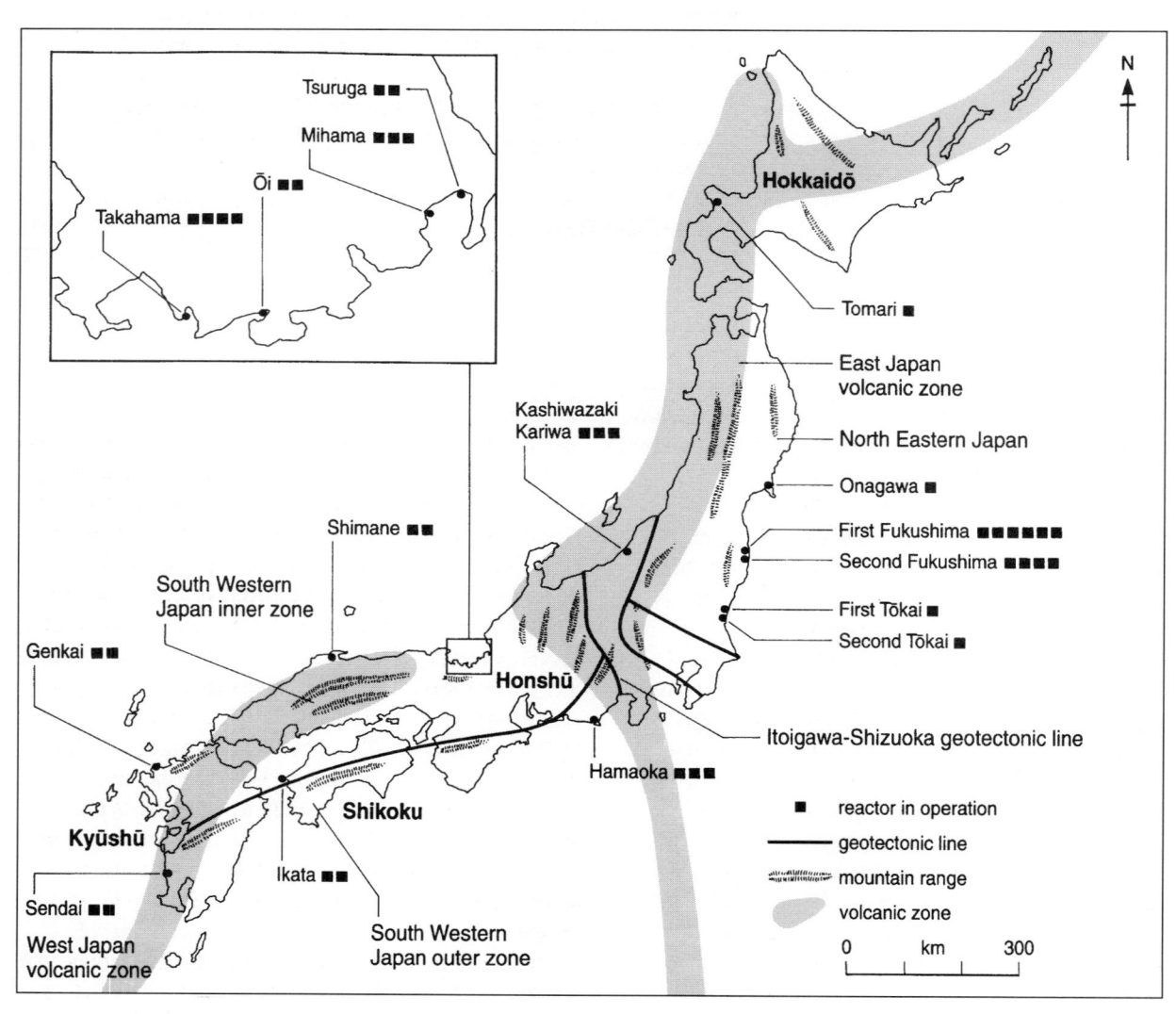

Figure 3.7 *Nuclear power and the geotectonics of the Japanese Archipelago*

SECTION 4

Africa: energy rich and energy poor

KEY IDEAS

- **Energy wealth alone will not necessarily assist a country in its development. Investment in other sectors of the economy and factors such as exchange mechanisms, efficiency in government, and economic policy are as important as initial energy wealth.**
- **In East African countries, the continued use of 'rural' energy fuels in the rapidly growing urban areas reflects a lack of economic development.**
- **The reliability of commercial energy supplies reflects the economic and political structures of a society.**

NIGERIA AND THE SINATRA DOCTRINE

If Japan has few energy reserves then Nigeria represents the opposite situation. Nigeria's energy boom is related to the fortunes of its oil industry, and the significance of oil in the economy. Before oil was exploited, Nigeria had a rural society with an economy based on agriculture. Two attempts to explore for oil were made, and each abandoned because of the outbreak of the world wars. It was not until 1958 that oil was discovered and then exported. Initially Nigeria did not require the oil; at this stage in the country's development, energy requirements were modest. A coalfield at Enugu, and a small hydro-electric grid on the Jos Plateau were supplemented by the widespread use of fuelwood, some of which was even used to fuel small thermal power stations. The Kainji dam across the River Niger, completed in 1968, was the first project to bring large amounts of energy to the country.

Oil was first exported from Ogoniland and production was initially limited by the barge capacity of the fleet moving to Port Harcourt, but construction of a pipeline in 1960 permitted greater quantities to be moved. Until the end of the Biafran War in 1970, oil was only one export commodity amongst others: timber, rubber, cotton, and tin were the most notable. The successful cohesion of the nation after the war, the global rise of the oil industry and the trebling of the price received for oil after OPEC activity in 1973, all led to a rapid expansion of the oil industry as the graph in Figure 4.1 on page 34 demonstrates.

By 1982, 90 per cent of Nigeria's export revenues and, more worryingly, 82 per cent of the Government's income was provided by oil. Migration to the cities became the dominant population trend as people moved to find employment in construction-related jobs, in the rapidly expanding civil service which administered this money, and in the industries servicing the newly rich. As a result agriculture

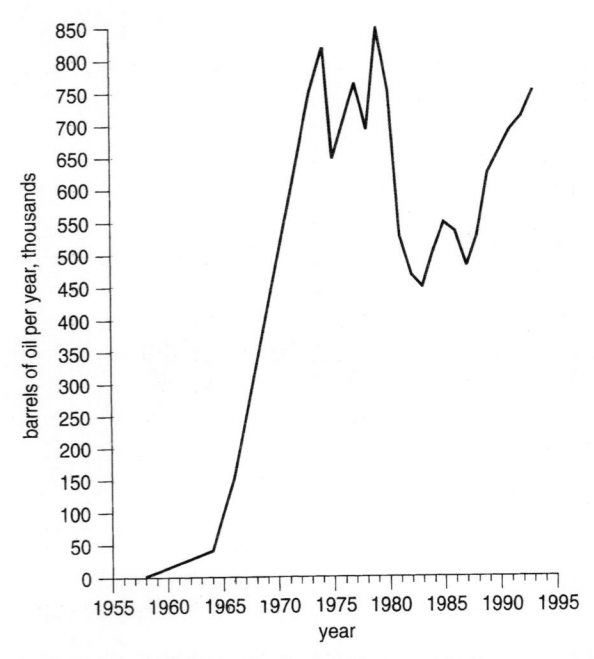

Figure 4.1 *Nigerian oil production since 1955*

suffered; exchange controls, which kept the value of the Naira high, made it as profitable to import food as to grow it. The new urban elite developed a taste for western food; imports of rice and wheat doubled during the 1970s. Manipulation of the Government spending programme, and there was much corruption became the route to wealth. Similarly, importing and the rapid turnover of goods, rather than their manufacture, built upon a well-established entrepreneurial tradition. Indeed in a society subject to political instability this made more economic sense to individuals than long-term investment. Industry, which by its nature requires sustained, long-term, slow-return investment, was neglected. Even in transport the pattern was biased towards consumption; the motorway network was expanded and the railways disregarded. Parallel to this, the demand for more regular supplies of petroleum products, coupled with the high tanker accident rate, led to the extension of the pipeline network. By 1980 3000 km of pipeline had been constructed and 19 storage depots were in use, three of them at refineries at Port Harcourt, Warri and Kaduna.

One of the main positive effects of the oil boom, however, was the investment in education. Primary enrolment tripled, and 20 new universities were built. As recently as 1985, more than 100 000 Nigerian families were sending one or more children abroad to university. Tribal divisions became less significant in a society dominated by a new urban educated bourgeoisie.

The oil shock of 1973 and the global cut-back of industry exposed the weakness of the Nigerian economy (see Section 1). Many factories had to close, for they relied on imported, rather than local materials, and by 1978 foreign loans had become necessary. Imports were restricted; Government expenditure was cut; and work on the new federal capital at Abuja was stopped. The Iranian revolution gave some respite with the cut of petroleum supplies, but this did not last. By 1982, the Government introduced a price cut below the agreed OPEC limit to try and expand markets and thus revenues. No amount of economic tricks, however, could conceal an economy in distress. The 1980s saw a saga of debt-rescheduling and hesitation. For example, the Shiroro HEP scheme on the River Kaduna (Figure 4.2) was put on hold, just short of completion, between 1982 and 1991. Loans from the EU to restructure and invest in agriculture; stringent measures against corruption in the waste disposal industries; and plans to channel resources to small businesses suggested that Nigeria was facing up to its difficulties. Nevertheless, as the oil boom has declined, old tribal and religious rivalries have re-emerged. In clashes between Muslim and Christian groups 246 people died in April 1991. Abuja, a symbol of social and religious cohesion, remains unfinished.

Petroleum and its products remain at the heart of the nation's economy and psyche. Ogoniland's quest for social justice is linked to the environmental degradation which the oil industry has brought. Greenpeace estimates that 40 per cent of all Shell's world-wide oil spills between 1982 and 1992 were in the Ogoniland area. Petroleum shortages in 1993 resulting from an outbreak of fire at the Kaduna refinery; the closure of the Warri refinery; increased smuggling of products to neighbouring countries; and strikes by tanker drivers; led to a rise in prices. The Government revoked export permits, and planned to introduce a new gasoline at, of course, greatly enhanced prices. This led to a nationwide strike by members of the trade union, the Nigerian Labour Congress. The power of the labour unions was largely arrested in August 1994, and in October all categories of fuel were tripled in price by the military junta. Allegations that $12 billion earned from the petroleum industry had disappeared between 1990 and 1994 whilst in transit to 'special accounts' opened by the Government abroad, did little to discourage corruption at the local scale. For example, during most of 1994, 'prospectors' sold fuel tapped from a leaking pipeline at Lagos airport, their only costs being bribes to the police.

In 1991, coal was shipped out of Nigeria for the first time in 20 years from the Eagle mine at Enugu, suggesting that other energy sources were being reassessed. Most Nigerian coal is high-grade, non-coking coal and unsuited to industrial processes and so it is unlikely that the high capital costs of mining and **beneficiation** and the low world prices for equivalent quality coal will allow for much expansion of the mining industry.

The electricity industry also reflects a growing crisis. Supplies are still typical of less developed societies, with demand not being met. An advisor to investors observes 'No company, and few wealthy Nigerians can avoid investing in independent generating capacity'. Overall capacity installed by National Electric Power (NEPA) was 3272 mw in 1991, an increase of 11 per cent over the previous year, due to the commissioning of a 600 mw gas turbine plant in Delta state. Consumption in 1993 was 24.9 per cent for commerce and street lighting, 24.9 per cent for residential use, and 50.2 per cent for industry.

Every country deserves the right to 'do it my way'. Little reference was made to external independent advice in the case of Nigeria and the temptation to spend rather than invest the oil revenues proved irresistible. Critics claim that North Sea oil revenues were similarly spent. The 'Sinatra Doctrine' applies to developed and less developed societies alike. Writing at the time of the greatest excesses of the spree, the Nigerian poet, Aig Higo issued a warning.

It remains to be seen what the full consequences of ignoring this shall be.

I struck tomorrow in the face.
Yesterday groaned and said,
'Please mind your steps today'.

Hidesong, *Aig Higo*

One of Nigeria's most valuable energy resources is of course its people, with a tradition of enterprise and high levels of education. The significance of Nigeria as a voice for Africa and as a model of how mistakes in the development process could be avoided should not be underestimated.

Q47. *Refer back to Figure 1.4 on page 7. How does the above account help explain why the figures for oil consumption in Nigeria are 'not available'?*

Q48. *How could investment of oil revenue in education abroad for some people have both positive and negative effects on a country like Nigeria?*

Q49. *Figure 4.2 shows a map of oil pipelines and power stations in Nigeria. Describe the regional distribution of these energy installations. Do they correlate with the centres of population?*

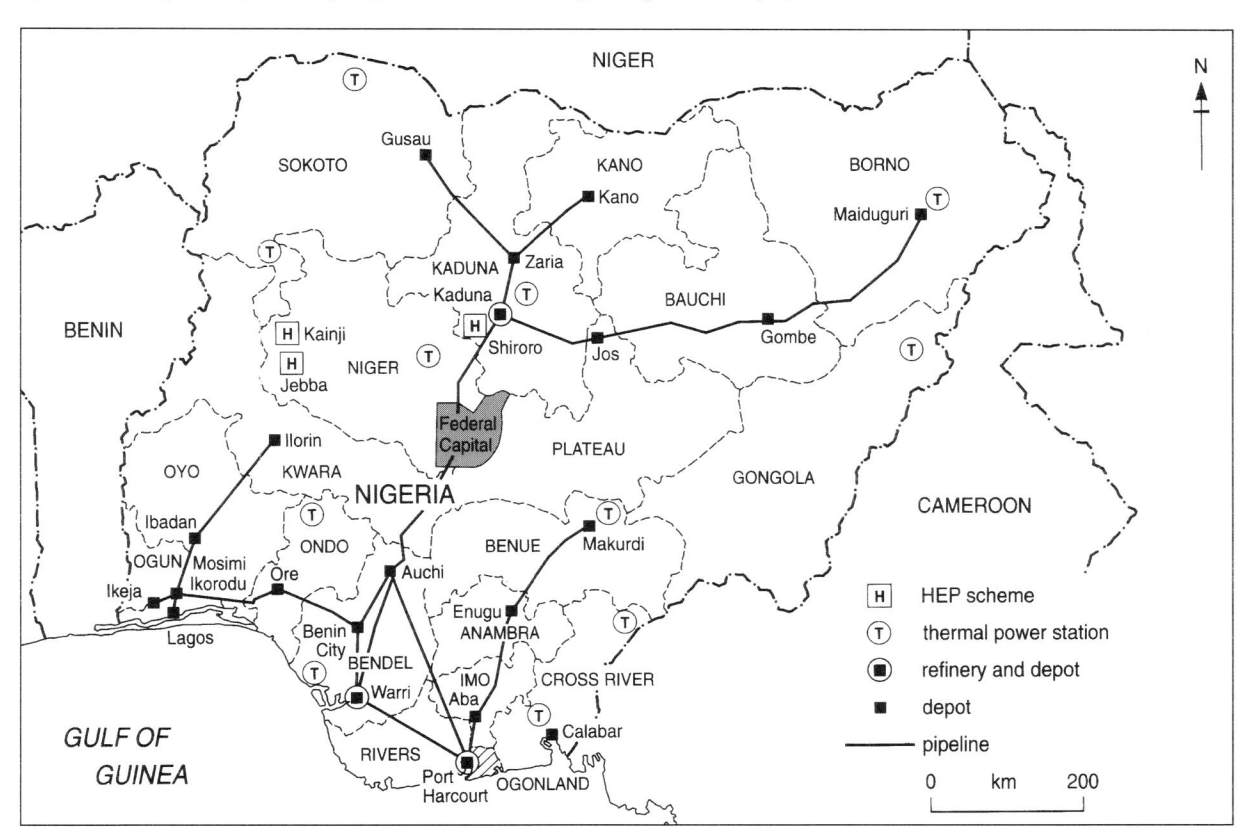

Figure 4.2 *Oil installations and power stations in Nigeria*

EAST AFRICA: AN ENERGY CHALLENGE

Shifts of power since the end of the Cold War have done little to alter the status of East Africa. Burdened by high debt, low international trade, high birth rates, low life expectancy, low calorie consumption, and local conflicts, the region remains marginal to global strategic and economic interests. Among the major regions of the world this area, from the Sudan to Zimbabwe, has the lowest proportion of its inhabitants living in towns and cities. It is also one of the poorest regions of the world. The combination of these two factors means that the bulk of energy demand in these countries is for cheap energy, in the form of biomass. Firewood, charcoal and animal dung are, of course, only renewable if a cycle of tree regeneration and stock replacement occurs (Figure 4.4). The World Bank, basing its work on Tigray, Ethiopia, has drawn attention to the pressures that rapid population growth and a growing urban demand for fuel can make on the land.

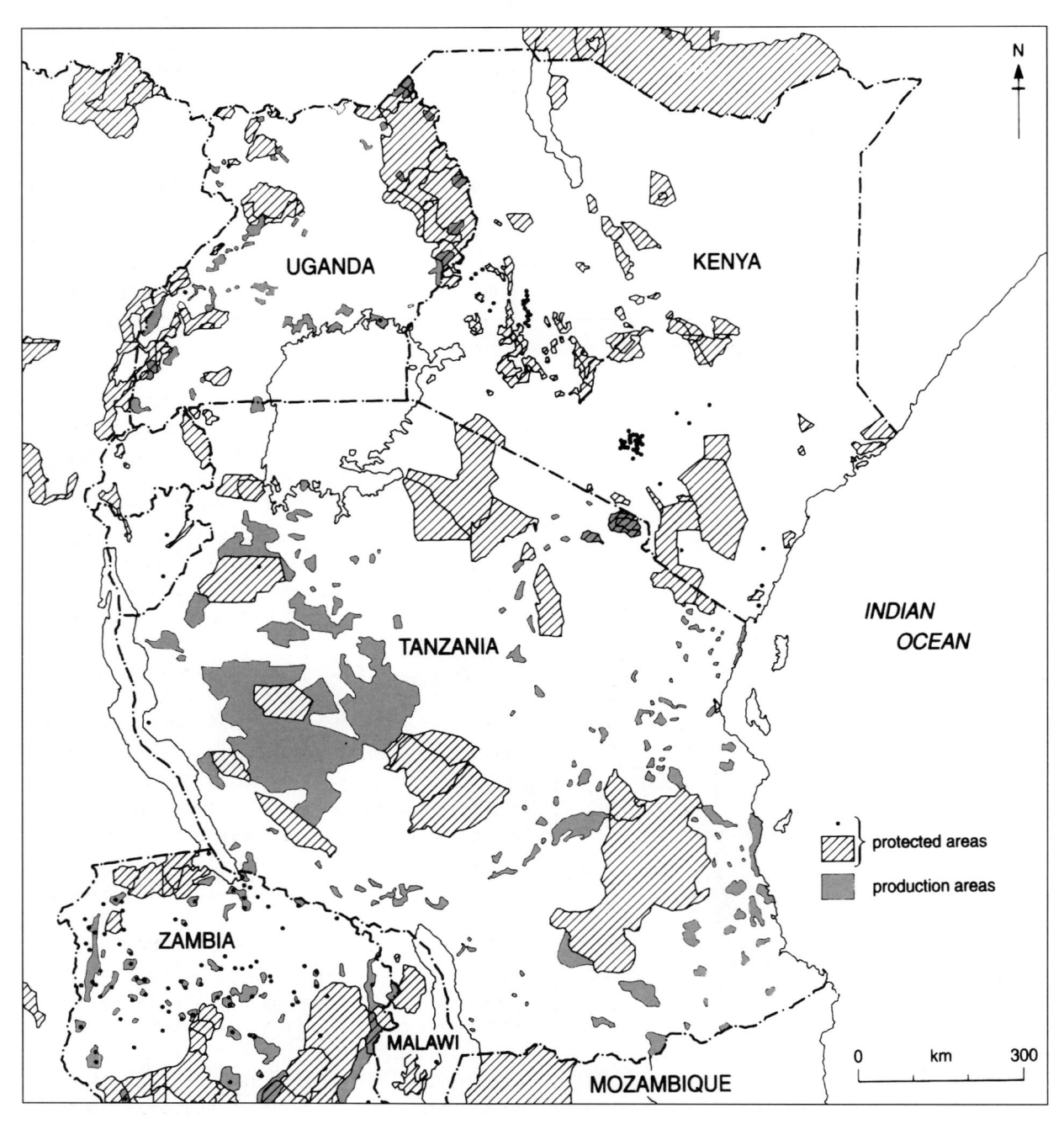

Figure 4.3 *Environmentally protected areas and forest production areas in East Africa*

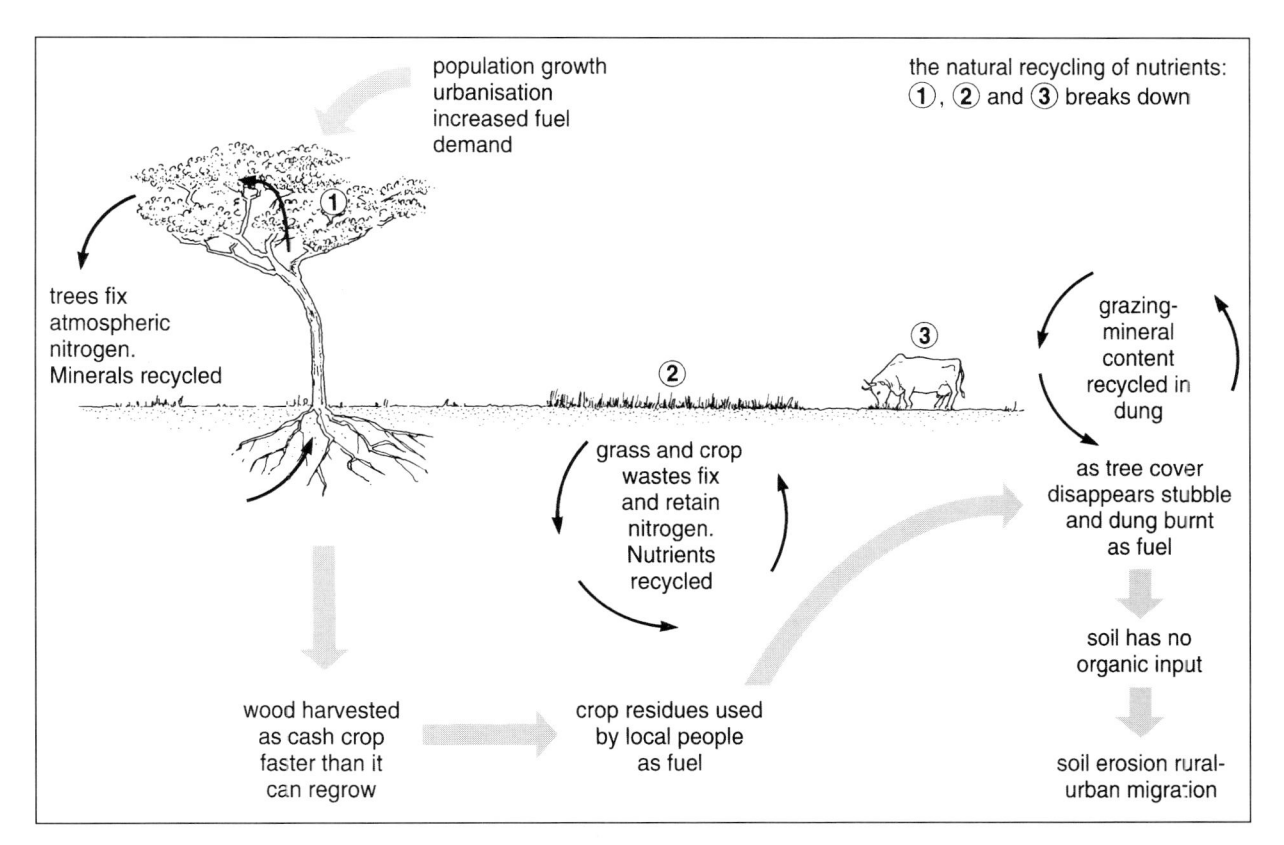

Figure 4.4 *The cycle of firewood related degradation in the environment*

Figure 4.5 *Location map of East Africa*

It would be simplistic, however, to accept that **over-population** is the sole cause of this pressure for firewood. A village study in central east Sudan showed journey times to adequate firewood supplies had doubled in 20 years. This was because of clearance of the land for irrigated cultivation of cotton and groundnuts in the Suki Agricultural Project. Changing land uses and agricultural economics, then, may put even more pressure on existing areas of woodland.

There are schemes to supply cities from peri-urban tree plantations, for example, at Kkibaha outside Dar-es-Salem. Wood from these commercial plantations is, however, very expensive. In contrast, rural wood-fuel supplies are usually cut or gathered from trees in common ground. The only cost is the labourer's time and the transport. The price is sufficiently rewarding to encourage firewood collection, but not high enough to stimulate planting and commercial growing.

Transport costs are kept low by using casual means of movement where possible. Wood is stacked by the main roads and sold to truck drivers, often carrying other loads, who will resell the bundles in the towns. Charcoal, with a better weight/energy ratio, is often transported by more formal arrangements. Thus road links, rather than radiating commercial energy consumption patterns

AFRICA: ENERGY RICH AND ENERGY POOR **37**

Figure 4.6 *Many new stove designs have been introduced in East Africa, often by aid agencies*

from the cities to the rural areas, are often used in East Africa to allow rural energy consumption patterns to continue in the cities.

Other attempts to reduce wood-fuel consumption focus on conservation of fuel resources at the point of consumption.

Urbanisation is a relatively recent phenomenon but it is occurring quickly due to migration and rapid population growth in all East African countries as Figure 4.7 illustrates.

Figure 4.7 *Urban trends for selected countries in East Africa*

COUNTRY	POPULATION GROWTH DURING 1980s (%)		LARGEST CITY: % OF TOTAL POPULATION
	TOTAL	URBAN	
ETHIOPIA	2.9	5.2	37
KENYA	3.8	8.2	57
MALAWI	3.4	7.9	19
MOZAMBIQUE	2.7	11.0	83
SOMALIA	3.0	5.6	34
SUDAN	3.1	4.1	31
TANZANIA	3.5	11.6	50
UGANDA	3.2	5.1	52
ZAMBIA	3.7	6.7	35
ZIMBABWE	3.2	6.2	50

The urban systems in these countries are clearly dominated by the **primate cities**. Kenya attempted to stop this trend with its second National Development Plan, 1970–1974, while Tanzania tried to relocate its capital, unsuccessfully, to Dodoma. Malawi has recently adopted a growth centre strategy. Nevertheless, growth of the largest cities continues. It is important to recognise that this process of urbanisation is proceeding without industrialisation. In the smaller cities particularly, biomass fuel sources as opposed to commercial ones are still crucial to households.

Why is there not a greater shift to modern, commercial energy fuels in the urban areas? The first reason is that these countries have limited economic resources. The sector with money to spend on energy – the urban middle class – is very small and a high proportion of earnings may be sent back to the extended family in the rural hinterlands. The second reason is more practical: the supply of most modern energy supplies is less reliable than that of charcoal or firewood. Kerosene and liquid petroleum gas are not always obtainable, and electricity systems experience regular cuts. Problems of reliability can be attributed in some countries to military activities and the security situation, interfering with regular supplies. The electricity utilities are also weak. Illegal connections to shanty areas may tap a large proportion of the supply. There is difficulty in collecting bills and disconnecting non-payers, particularly in countries such as Tanzania and Somalia where the largest non-paying customers are government ministries! The lack of industrial consumers means that the electricity systems are dominated by households, with expensive consumption patterns. Two diurnal peaks, one at mid-day and one in the evening are expensive to supply.

Another reason for the lack of reliability is that the full cost of electricity is rarely met by the customers. In Tanzania, for example, a 'life-line subsidy' of one shilling per kw/h for the first 1000 kw/h used each month provides the average family with most of their lighting, cooking and air conditioning needs. It is too often accepted that the cost of cheap electricity is frequent blackouts. Consequently, even where incomes would permit it, the fear of power failures means there is a reluctance to switch entirely from firewood to electricity for food preparation in the cities.

It is strange that in most East African societies, market forces determine the supply and cost of firewood and charcoal, yet there is a resistance to charging the full market price for electricity. In East Africa urbanisation is proceeding without a parallel growth in industry, and there is a slow change in domestic energy use. Biomass is still a dominant energy source, and its use has profound implications for the rural environment. A move towards greater use of electricity could, however, because of its expense, place pressure on these countries already burdened by the cost of importing fuel to fire power stations. Many areas of East Africa are ideally suited to the development of hydro-electricity but this can only be an answer if there is massive investment from outside. Photo-voltaic technology remains commercially immature and untried except in small-scale projects in the region.

Many projects assisted by outside agencies aim for self-sufficiency in energy. Typical is Chogoria 'bush hospital' in central Kenya which uses both solar panels and firewood cut from the 'pin oak' (*Grevillea Robustica*). This tree regenerates quickly, and its tap roots allow it to be planted close to other plants.

Q50. a) Why is commercial fuelwood expensive?
b) Refer to Figure 4.3 on page 36. What would be the implications of extending the protected areas?

Q51. Use an atlas to name the largest city in each country shown in Figure 4.7. Suggest why the statistics from other countries in East Africa may have been less reliable, and so not included in the table.

Q52. Why can few countries in East Africa accommodate variations in daily demand in the way that Britain provides peak and baseload? (Refer back to Section 3).

Q53. Outline the reasons for a continuing reliance on biomass energy supplies, rather than commercial energy supplies, in East African cities.

Q54. Referring back to the case study of Nigeria (pages 33–5) consider the implications if economic circumstances in East Africa were to improve. Would energy supplies necessarily improve? Would energy consumption patterns change?

Q55. Why are many projects, such as hospitals and schools, self-sufficient in energy?

Figure 4.8 *Cahora Bassa hydro-electric dam in Songo Mozambique*

DECISION MAKING EXERCISE

An energy policy for Newfoundland

Canada is a classic case of **regional disparity**. If Ontario and the Vancouver area are the dynamic, wealth-creating **cores**, Newfoundland is the archetypal **periphery**. It is perceived by many Canadians as a distant, but persistent drain on the country's wealth.

The thin acidic soils on ice-scoured rocks of the Canadian shield, and the climate, militate against agriculture. The province has always relied heavily on fishing and forestry. The 1990s have seen a dramatic decline in fish stocks, and a complete collapse of the cod industry. Boosting the economy with tourism is difficult. The rise in temperature in spring melts pack-ice to the north. Ice-bergs then flow south, bringing fogs and cool weather in the early summer. In addition, the cold Labrador current suppresses temperatures throughout the year.

Newfoundland, with a population of 548 475 (including 30 375 in Labrador) is characterised by isolation, low industrial development, and a high level of out-migration. Unemployment was 34.1 per cent in March 1993 compared with a national average of 11.2 per cent. Many of the problems are typical of remote communities. In 1993, the average earned income of $11 499 was 37 per cent lower than the national average of $18 155. Even average income including state benefits of $17 448 was 21 per cent lower than the national average.

Newfoundland's recent energy crisis reflects the island's economic status. Early in 1994 the Minister had to limit the casual harvesting of firewood as the failure of the fish industry encouraged unemployed fishers to cut increasing quantities of wood to burn, or barter. Newfoundland is a high consumer of firewood; up to 25 per cent of some forested areas are cropped for domestic fuel. Recession and the failure of industry have intensified the need to save on oil and electrical energy.

An exploration permit was issued to Mobil Oil in 1965, but the Hibernia field was not discovered until

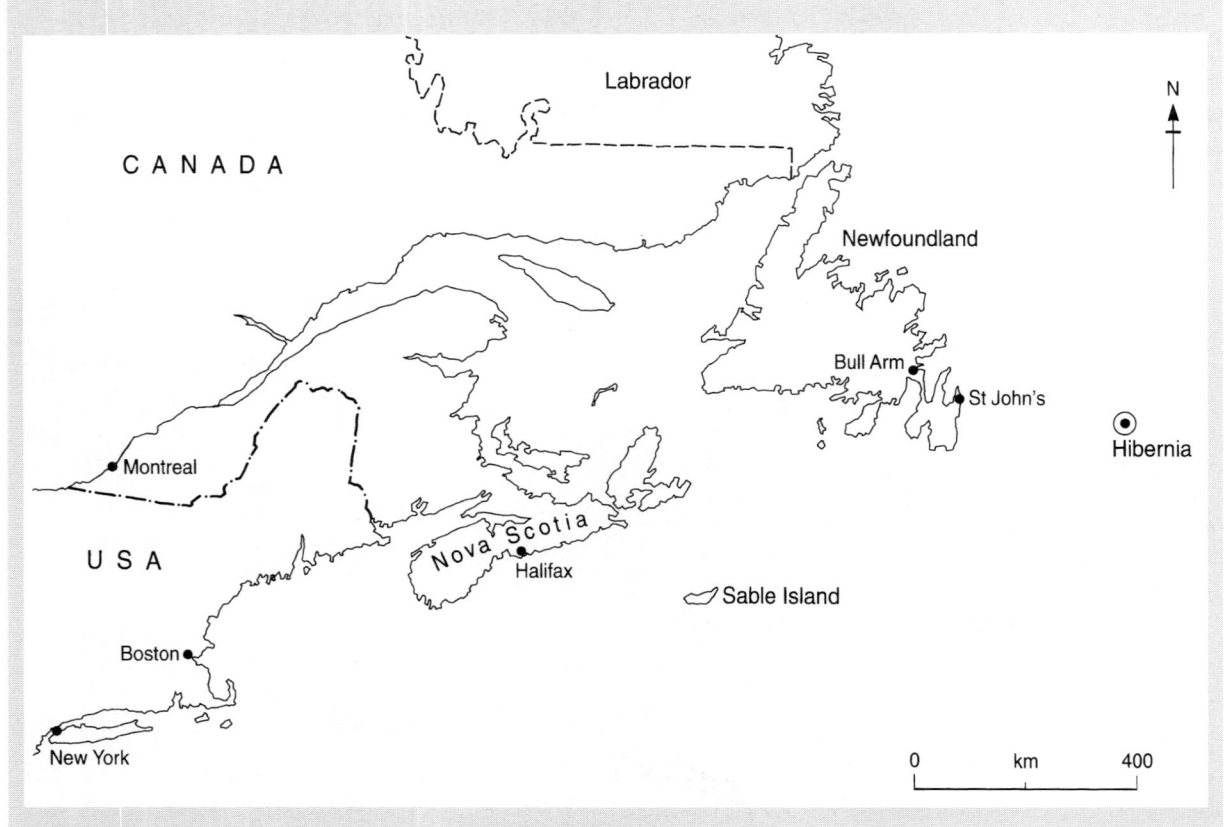

Figure 5.1 *The location of Newfoundland and the Hibernia field*

40 DECISION MAKING EXERCISE

1979, and only declared a commercial discovery in 1990. Delays were caused by the dispute in ownership of the field, resolved in the Atlantic Accord, signed in 1985. This established Federal and Provincial co-operation and drew up a management framework. Newfoundland and Labrador stand to share half the profits, the commercial companies receiving the rest. A full geological description was also time-consuming and costly: the nature of the reservoirs made it difficult to estimate the quantities of recoverable oil and the number of wells needed to obtain it.

Mobil's exploratory drilling confirmed the existence of major hydrocarbon-bearing sandstone units in sediments of Jurassic and Lower Cretaceous age. The Jeanne d'Arc zone, in the Jurassic sands, and the Hibernia, Avalon and Seismic B-Marker zones in the lower Cretaceous sand, have been named. These multiple sand layers are non-communicating and are within a complex folded and faulted structure; a consequence of tectonic and depositional processes. Large faults bound this roughly triangular area of 85 km^2. Both the Avalon and Hibernia zones can be sub-divided into three entities: the upper, main, and lower sands. In the Hibernia field the main sands represent the primary reservoir, and thirteen soil horizons, each separated by shale barriers, can be identified. Recovery of oil will be by using a combination of waterflood and reinjection of produced gas techniques. The Avalon sandstones are more stratified and have lower reservoir quality than the Hibernia layers, so production from this zone is not planned until 2005. Both the Hibernia and Avalon reservoirs will be reached with 83 wells connected to the production platform, the Gravity Base Structure (GBS).

Hibernia crude oil is of similar density to West African crudes, with a sulphur content by weight of 0.4 to 0.6 per cent. It could therefore be sent to similar markets on the US East and Gulf Coasts, and in Europe. With a pour point of 12 °C, local refining could incur storage and handling costs. The 1990 estimate of recoverable oil was for 540–650 million barrels. With average production of 110 000 barrels per day, the life of the field is estimated at 18 years.

Obtaining the oil from such a challenging environment is, however, expensive and has already demanded original design solutions, outlined in Figures 5.3 and 5.4 (page 43).

Environmental challenges to the oil industry in Hibernia

Sea floor characteristics
Water depth ranges from 79–84 m. Sea floor slopes are less than 0.8 per cent (0.5 degrees).

Geological processes
There is no active faulting at the site and there are low seismic excitation levels. The ocean floor material is granular and cohesive so that pile foundations, the Gravity Base Structure (GBS) and the pipeline would be feasible.

Oceanography
Based on a 100-year return period, extreme currents range from 1.69 m per second at the surface to 0.66 m per second near the sea floor. Maximum wave height for a similar return period is 29.3 m. The tidal range at Hibernia is estimated to be 1 m at the maximum, and the extreme storm surge 0.7 m. Extreme surface sea water temperatures range from −2.5 to 15.6 °C.

Meteorology
Monthly mean air temperatures at Hibernia range from −0.5 °C in February to 13.5 °C in August.

Monthly mean air temperatures at St Johns range from −4.5 °C in February to 15.5 °C in July.

Precipitation at St John's, considered to be more extreme than at Hibernia, has a greatest recorded rainfall of 121 mm in 24 hours and 55 cm of snowfall in 24 hours. On average precipitation falls as snow on 85 days per year.

Maximum wind speeds at an elevation of 10 m based on a return period of 100 years are 34.5 m per second average over 1 hour, with gusts of up to 49.3 m per second.

Visibility is best in fall, poorest in early summer. The poorest flying conditions are during fogs of the late winter and early spring.

Icing has not been significant on floating exploration drilling units. The highest probability of icing exists in February, when severe icing is estimated to occur almost 2 per cent of the time.

Storms rarely occur during the iceberg season of May–June. There is low probability of a severe storm coinciding with iceberg impact.

Ice and icebergs

Sea ice has drifted into the Hibernia area during six of the 27 winters between 1959 and 1985. Average ice concentration increases from early January to a maximum in mid-March. Ice covers the whole area only in extreme conditions. Mean duration of ice has been two to three weeks, the maximum, six weeks.

An annual average of 30 icebergs pass through the degree square containing the Hibernia field. Most drift in May–June. The general drift of icebergs, at 0.2–1.0 m per second, is easily monitored but short-term local behaviour is difficult to predict. The largest iceberg mass which could drift into the Hibernia area has been calculated as approximately 6×10^6 tonnes.

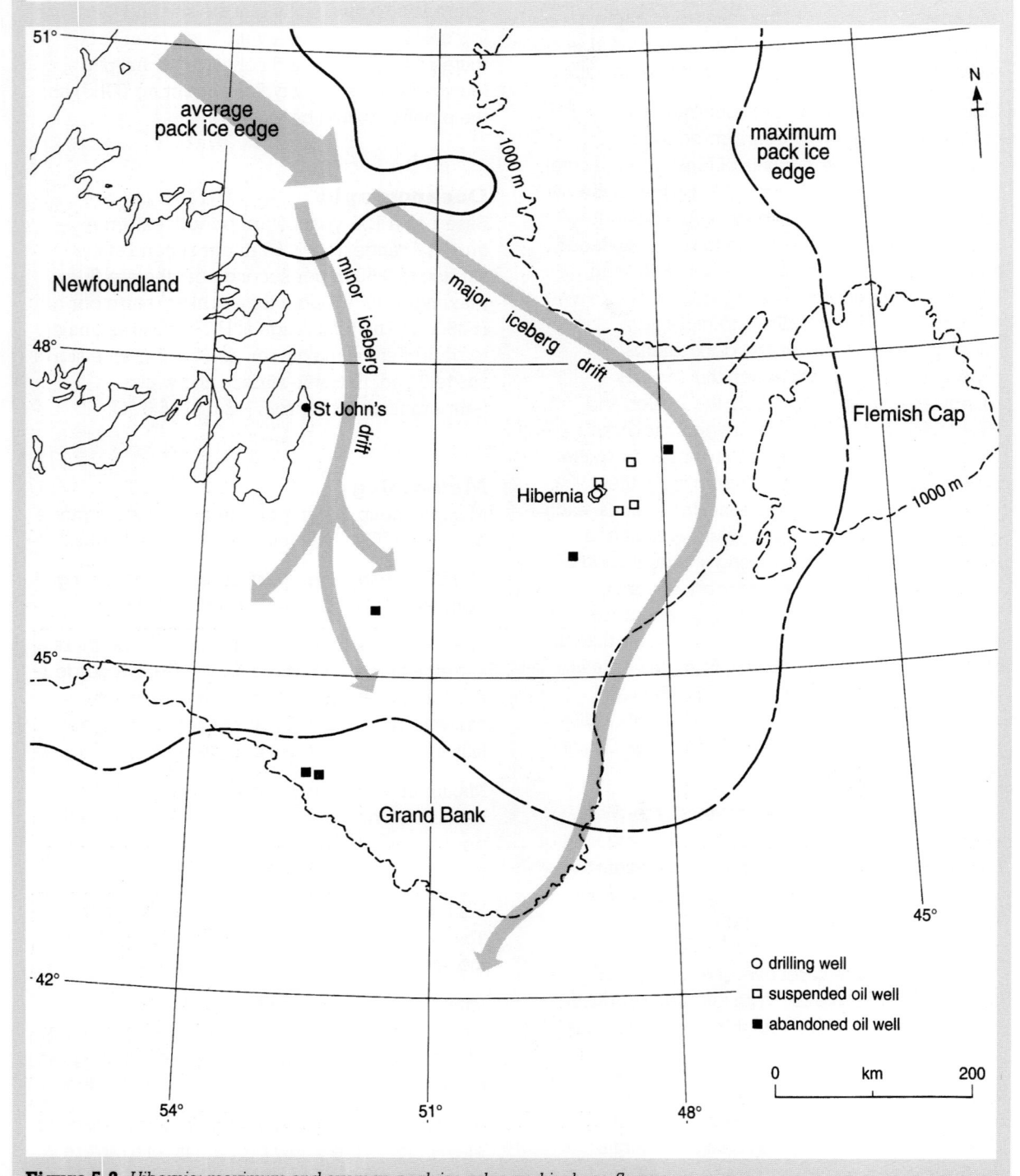

Figure 5.2 *Hibernia: maximum and average pack ice edge and iceberg flows*

Design solutions

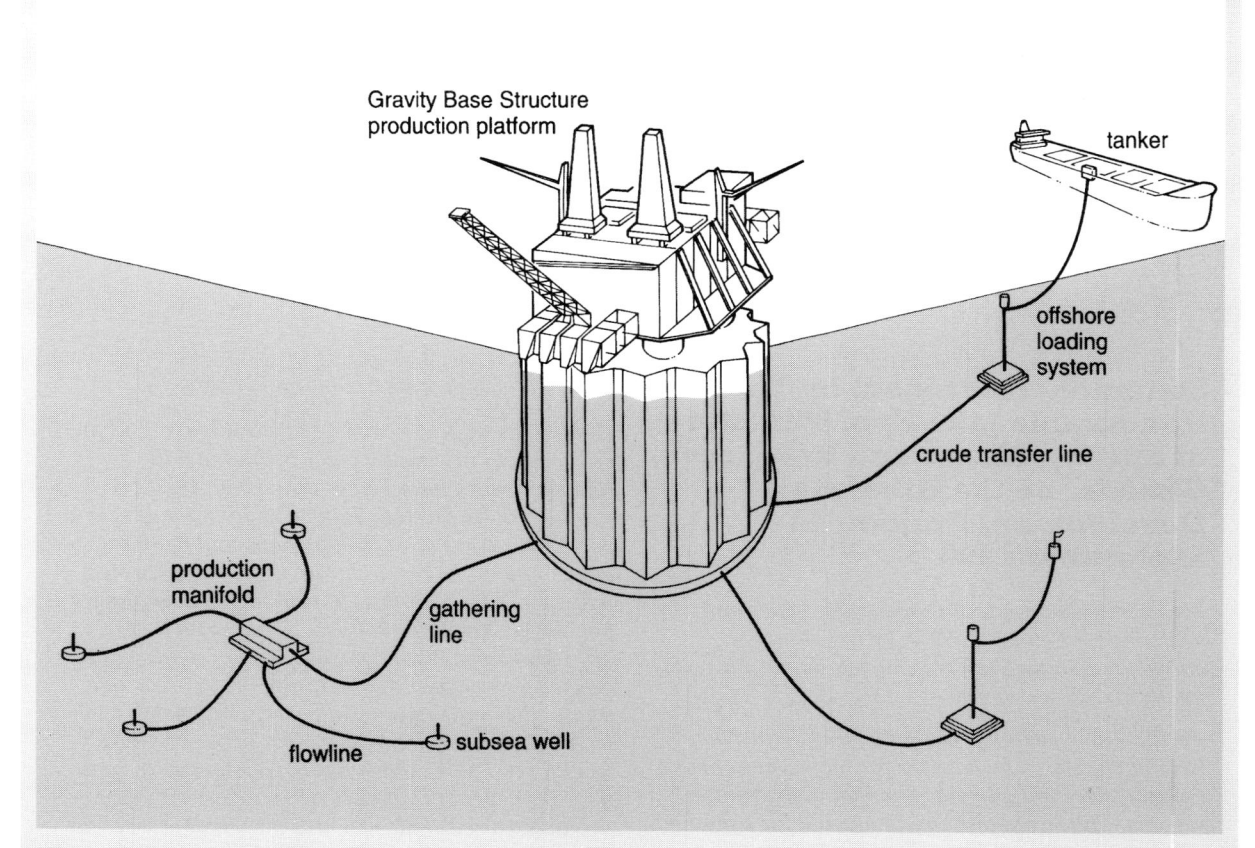

Figure 5.3 *Production system components*

1 module transportation

2 topsides assembly

3 completed topsides

4 load out

5 deck mating operation

6 tow to Hibernia Field

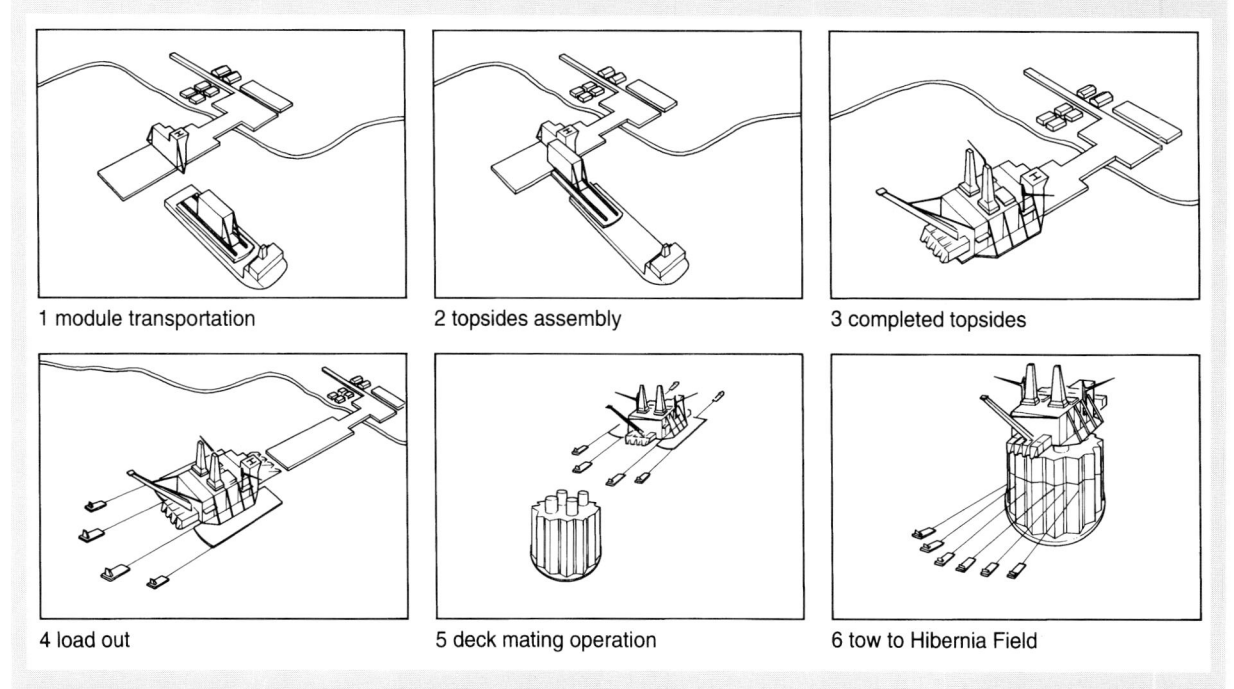

Figure 5.4 *Topsides construction and assembly*

As the diagrams in Figures 5.3 and 5.4 on page 43 show, principal components of the production system are the GBS, the topsides, the subsea installations, the offshore loading system and the crude oil transport tankers.

Three tankers of 120 000 dead weight tonnes (DWT) will each carry 700 000 barrels of oil. Double hulled and ice-reinforced they will be supported by ice clearing and support vessels. Other technical details are listed below.

Construction of the GBS, assembly of the completed topsides and mating of the topsides with the GBS will take place at Great Mosquito Cove, in Bull Arm, Trinity Bay, 150 km north west of St John's near the communities of Sunnyside and Come by Chance.

Total height of GBS: 213.1 m;
total weight under tow: 600 000 tonnes;
total weight after ballasting with iron ore: 1 million tonnes.

Statement (abridged) by The Honourable Jake Epp, Minister of Energy, Mines and Resources, Canada, on the Hibernia Development Project, 14 September 1990

Mr Premier, honourable colleagues, ladies and gentlemen; I am delighted to be here today to announce that the Hibernia Development Project is finally a reality.

Today a new energy province is born. Hibernia heralds the start of a new era for the development of Canada's vast oil and gas potential in its frontier regions. This is a major breakthrough for Canada's oil and gas industry and will contribute to the ongoing diversification of Canada's energy supply sources.

The effects of this development will be long-lasting, not only for Newfoundland, but for all of Canada. Hibernia is only the first step towards the new skills, expertise and technologies required to meet the challenge of frontier development. A modern infrastructure and workforce will set the stage for future oil projects, such as Terra Nova and Ben Nevis, and will place Canada at the leading edge of offshore technologies.

We are doing no less than confirming the long-awaited economic renewal of the Province of Newfoundland and Labrador, and our faith in Canada's future. All environmental assessments have been completed and the Government of Canada is satisfied that Hibernia can now proceed. The Atlantic Accord of 1985 cleared the way for the project. As was outlined in the Statement of Principles, signed in 1988, the Government of Canada will provide the consortium with a cash contribution for 25 per cent of eligible pre-production capital costs, up to a maximum of $1.04 billion. We will also guarantee loans for 40 per cent of those costs to a maximum of $1.66 billion. In sum, the Government of Canada will invest a total of $2.7 billion to make Hibernia a reality.

Getting the oil to the surface is a challenge. But the rewards will be equally great. Hibernia will represent more than 12 per cent of our light oil production by the year 2000. Newfoundland's economy, at long last, will be bolstered by this development. Every dollar invested in this project will yield several dollars in primary and secondary benefits to Newfoundland. It is estimated that approximately 10 000 person years of work will be created in Newfoundland during the construction phase. From the long-term perspective, new technical skills will be developed in the province's workforce – skills that will take advantage of future developments in Canada's offshore areas. These newly acquired skills will also be valuable assets in developing and diversifying other sectors of the Newfoundland economy.

The Governments of Canada, Newfoundland and Labrador have set a regional benefits package for Newfoundland and Labrador. As a consequence, the Hibernia consortium has agreed to fabricate one of the super modules in Newfoundland, and will use its best efforts to build a second super module in Canada. It is expected that the total Canadian participation in the Hibernia project will equal 55 to 60 per cent of project costs. This compares favourably to the first North Sea projects, which achieved around 25 to 35 per cent domestic content. Firms across all of Canada will have the opportunity to participate, thus creating real jobs.

Through the joint initiative of two governments and the private sector, the enormous economic benefits from Hibernia are now upon us.

Long before I entered politics I was a teacher of history in my home province of Manitoba. I know all too well that history judges, with the benefit of perfect hindsight, whether the course chosen was the best one. I am confident that history will view our decision here today as a fine example of the foresight and determination of which great countries are made.

Instructions

Read all of the information in this section.

1. Write a report outlining the difficulties faced by Mobil in exploiting the Hibernia field.
2. During what season of the year would you recommend the towing of the Gravity Base Structure to the well site? Why?
3. With a partner, attempt to outline the objections to, and the motives for, exploiting the oil. Imagine that you work for the Ministry of Energy and prepare a report on why you would support or oppose the Hibernia Development Project. Finally, considering your own views, why do you think that the decision to exploit the oil was finally favoured?
4. Read the Minister of Energy's speech. What lessons would you draw from the 'North Sea Bubble' to advise Newfoundland on how to prepare for the rise and decline of the oil industry?

PROJECT SUGGESTIONS

1 Evaluating power station sites

a) Using maps of various scales, gather evidence to propose suitable sites for a gas-fired power station in the UK. Identify five areas of relatively flat, open and stable land that are near urban markets. Create a spreadsheet and for each site evaluate the following criteria on a scale of 1–5 (1 = excellent, 5 = poor):
- distance to a gas terminal or pipeline (see Figure 2.5, page 17);
- access to water supplies;
- visual impact (including the screening potential of nearby higher land or of local planting);
- environmental impact (include factors of isolation and nearness to settlement, the building of access roads, placing of high voltage transmission lines and prevailing wind direction related to noise and effluents);
- status of the site (closeness to other industrial sites, to derelict land or conservation sites).

b) Assess the suitability of a rural site for a wind turbine. Through study of this book and other sources of information on wind power and conservation sites, identify factors considered when siting a wind turbine and factors contributing to high landscape quality. The latter might include distance from roads and built-up areas, vegetation type, height, slope, etc. Create a 5-point scale for recording these factors in relation to each site. Identify a number of potential sites in an area. Devise a questionnaire or check list to measure either people's perception of the potential sites on each factor or an objective measure of these factors for each site. Analyse the findings and reach a conclusion about the best site.

2 Measuring air quality

Lichen are particularly sensitive to air quality. Providing other factors are controlled, lichen types reflect the types of air pollution. Using a 1:25 000 map of a nearby urban area, identify sample sites at regular intervals, e.g. grid crossings. At each site identify a building or wall of a certain material or a tree trunk and the direction it faces, e.g. North. Record the location and the pollution characteristics of the site, as in part a) above. Then measure and record the types and proportion of the surface covered with lichen. A suitable instrument would be a 0.5 m quadrat divided into 5 cm squares, each representing 1 per cent of the area.

Map and/or graph the lichen cover. Relate the patterns found to potential sources of pollution such as various volumes of traffic, road junctions, industrial or other effluents, etc.

Further information on lichens may be found in Richardson DHS, Pollution Monitoring with Lichens, Richmond Publishing Co Ltd.

GLOSSARY

Acid flushes a sudden increase in acidity levels in a catchment area after a precipitation event of high acidity.

Acid rain precipitation polluted with most notably sulphur dioxide and nitrogen oxide.

Active solar device equipment designed to convert solar energy to heat and/or light, e.g. solar panels.

Alternative renewable source renewable energy derived from sources not used recently on a commercial scale in developed societies.

Anaerobic digestion a heat-producing process whereby waste is consumed by bacteria which do not use oxygen.

Associated gas gas found in the same location as oil.

Base load demand for electricity which remains steady throughout the day.

Beneficiation treatment of a raw material to improve its qualities for a particular purpose.

Biomass living matter.

Biosphere areas of the earth's crust and atmosphere occupied by living matter.

Capital-intensive a project requiring high financial investment.

CIS Commonwealth of Independent States (former USSR).

Clean Air Act legislation in 1956 by the British Government to improve air quality.

Commercial fuels fuels which are exchanged for money at a unit price, rather than bartered, and for which statistics are usually available.

Coolant a cooling material employed in a nuclear reactor.

Core a centre of economic growth attracting investment and migration.

Decommissioning the process of withdrawing a power station from use.

EC Thermie Programme European Commission fund to support alternative energy projects.

Electricity grid a national distribution system for electricity.

Fast breeder reactor a nuclear reactor which uses enriched uranium fuel to 'breed' plutonium and so intensify the fission process.

Fixed track system a system of distribution which requires investment in channels not easily moved, e.g. pipelines.

Geothermal heat heat generated within the earth's crust.

Greenhouse Effect the trapping of heat within the earth's atmosphere by gases, most notably carbon dioxide, which absorb radiation.

Groundwater water held at depth in rock reservoirs.

Isotopes forms of an element which differ in atomic weight but not in chemical properties.

Moderator a material used in nuclear reactors to delay and control the action of the neutrons.

Negative anomalies a below-average variable which does not conform to the general trend.

Negative externalities when a firm's actions result in a load or cost being borne outside the firm itself.

No-cost option a choice which observes certain principles without incurring extra costs.

Overpopulation a situation where there is an imbalance between the amount of people and the resources available.

Passive solar device a building design which conserves energy and maximises solar energy received.

Peak demand a time of maximum demand for electricity.

Periphery a region in a country characterised by economic decline, and out-flows of human and economic resources to the core.

Petro-dollars payment received, in dollars, for oil.

Primary biological activity the process of absorbing light energy by plants; photosynthesis.

Primary energy energy used directly, in the form it has been first obtained, e.g. oil burned as heating fuel.

Primate city the first city of a country usually, measured by population size referring only to the situation where it is many times larger than the second city.

Regional disparity uneven development and economic imbalance between parts of a country.

Retrofitting the redesign and adjustments made to power stations to reduce emissions.

Secondary energy energy which is transformed, usually into electricity, to make it more suitable for consumption, e.g. coal burnt to generate electricity.

Sequester to bind a chemical so that it cannot react.

Solar devices equipment designed to tap the sun's energy.

Totalitarianism a form of government which permits no opposition.

Tsunami 'tidal' wave.

USEFUL REFERENCES AND ADDRESSES

Atom Magazine (produced for the UK Atomic Energy Authority), AEA Technology, 329 Hartwell, Oxfordshire OX11 0RA

BP Statistical review of World Energy, The British Petroleum Company plc, Britannic House, 1 Finsbury Circus, London EC2M 7BA

Chell K, 'Landscape Assessment and Wind Power', *Geography Review*, Nov 1993

Chapman JD, *Geography and Energy*, 1989 Longman Scientific and Technical

Development of oil and gas resources of the UK 1992, Department of Trade and Industry, HMSO

Digest of UK Energy statistics, 1993, HMSO

Fairlie S, 'White Satanic Mills', *The Ecologist*, Vol 24 No 3, May/Jun 1994

The Electricity Council Public Relations Office, 30 Millbank, London WW1P 4RD

Friends of the Earth, 377 City Road, London EC14 1NA

Hanstock S, 'Global warming; accepting the implications', *Geography Review*, Mar 1991

Hirsch P, 'Power struggles on the Mekong', *Geography Review*, Jan 1993

Hosier R, 'Energy and environmental management in East Africa', *Cities Environment and Planning*, 1992 Vol 24

Hudson R, 'Coal resources and self-sufficiency; the case of Greece', *Geography Review*, Jan 1989

Hudson R and Sadler D, 'State policies and the changing geography of the coal industry in the UK in the 1980s and 1990s', *Transactions of the Institute of British Geographers*, Vol 15 No 4 1990

Mountfield PR, 'Electricity production after privatisation', *Geography*, No 329 Vol 75 Part 4, Oct 1990

Spooner D, 'Fossil fuels; the decade of surplus, Sid and Arthur', *Geography*, No 329 Vol 75 Part 4, Oct 1990

Twidell J, 'Renewable energy from sun, wind and waves', *Geographical Magazine*, Jul 1992

Policy Studies Institute, 100 Park Village, East London NW1 3SR

Renewable Energy Enquiries Bureau, ETSU, Building 156, Harwell, Oxfordshire OX11 0RA

Royal Commission on Environmental pollution, '17th report on Incineration of Waste', HMSO

Uranium Institute, Bowater House, 68 Knightsbridge, London SW1X 7LT

index

Acid rain 9
Air quality 45
Australia 23, 24, 30
Biofuels 24
Biomass 6, 24, 39
Brazil 24
Canada 22, 40–5
Carbon dioxide 8, 24, 27, 31
CFCs 8
Chernobyl 9, 20–1
Coal 4, 5, 12–14, 26, 34–5
Consumption 6, 7, 29
Denmark 28
East Africa 11, 36–9
Electricity grid 20
Energy flows 29
Energy issues 7–11
Former USSR 31
Fossil fuels 4, 8, 21, 22, 27
France 31
Gas 15–18, 26, 27, 30
Geothermal 20, 23–4
Germany 23
Global warming 7, 8
Greece 23, 24
Greenhouse Effect 8
HEP 22, 30, 34, 39
Hungary 23
Iceland 23
IPCC 8
Israel 23
Japan 30–2

National grid 26, 27
Netherlands 24
Newfoundland 40–5
New Zealand 22, 23
Nigeria 8, 33–5
North Sea 8, 15–18, 35
Norway 22, 25
Nuclear power 11, 18–21, 26, 27, 32
OECD 27
Oil 5, 15–18, 33–4, 35, 40–1
OPEC 7, 33
Peat 14–15
Philippines 32
Pollution 18, 25, 28, 31, 34
Power station sites 45
Privatisation 26, 27, 29
Production 6, 15
Radiation 18, 20, 21
Recycling 25
Renewables 21–5, 28, 36–7
Rio de Janeiro 8, 21
Scotland 9, 15
Solar 22–3, 32, 39
Switzerland 9, 23
Tidal 22
UK 11, 15, 24–5, 26–9
Uranium 6, 18–21
USA 9, 11, 24, 30, 31
Waste 24–5, 28–9
Waves 25
Wind 24, 27–8